POTENTIAL IMPACTS IN COASTAL ENVIRONMENT AND CLIMATE CHANGE

POTENTIAL IMPACTS IN COASTAL ENVIRONMENT AND CLIMATE CHANGE

S.K. Singh

RANDOM
RANDOM PUBLICATIONS
NEW DELHI - 110 002 (INDIA)

Potential Impacts in Coastal Environment and Climate Change

ISBN 978-93-51117-05-6

Published in 2015 in India by

RANDOM PUBLICATIONS

4376-A/4B, Gali Murari Lal, Ansari Road
New Delhi-110 002
Phone: +9111-43580356, 23289044
E-mail: randomexports@gmail.com; sales@randompublications.com; info@randompublications.com

Reprinted 2026

Type Setting by: Friends Media, Delhi-110089

Preface

The term "coastal zone" is a region where interaction of the sea and land processes occurs. Both the terms coast and coastal are often used to describe a geographic location or region; for example, New Zealand's West Coast, or the East and West Coasts of the United States. A pelagic coast refers to a coast which fronts the open ocean, as opposed to a more sheltered coast in a gulf or bay. A shore, on the other hand, can refer to parts of the land which adjoin any large body of water, including oceans (sea shore) and lakes (lake shore). Similarly, the somewhat related term "bank" refers to the land alongside or sloping down to a river (riverbank) or to a body of water smaller than a lake. "Bank" is also used in some parts of the world to refer to an artificial ridge of earth intended to retain the water of a river or pond; in other places this may be called a levee. While many scientific experts might agree on a common definition of the term "coast", the delineation of the extents of a coast differ according to jurisdiction, with many scientific and government authorities in various countries differing for economic and social policy reasons. According to the UN atlas, 44% of people live within 150 kilometres (93 miles) of the sea. Tides often determine the range over which sediment is deposited or eroded. Areas with high tidal ranges allow waves to reach farther up the shore, and areas with lower tidal ranges produce deprossosition at a smaller elevation interval. The tidal range is influenced by the size and shape of the coastline. Tides do not typically cause erosion by themselves; however, tidal bores can erode as the waves surge up river estuaries from the ocean.

The coast and its adjacent areas on and off shore are an important part of a local ecosystem: the mixture of fresh water and salt water in estuaries provides many nutrients for marine life. Salt marshes and beaches also support a diversity of plants, animals and insects crucial to the food chain. The high level of biodiversity creates a high level of biological activity, which has attracted human activity for thousands of years. The coast is a frontier that nations have typically defended against military invaders, smugglers and illegal migrants. Fixed coastal defences have long been erected in many nations and coastal countries typically have a navy and some form of coast guard. Coasts, especially those with beaches and warm water, attract tourists. In many island nations such as those of the Mediterranean, South Pacific and Caribbean, tourism is central to the economy. Coasts offer recreational activities such as swimming, fishing, surfing, boating, and sunbathing. Growth management can be a challenge for coastal local authorities who often struggle to provide the infrastructure required by new residents. Coasts also face many human-induced environmental impacts. The human influence on climate change is thought to contribute to an accelerated trend in sea level rise which threatens coastal habitats. Pollution can occur from a number of sources: garbage and industrial debris; the transportation of petroleum in tankers, increasing the probability of large oil spills; small oil spills created by large and small vessels, which flush bilge water into the ocean. Fishing has declined due to habitat degradation, overfishing, trawling, bycatch and climate change. Since the growth of global fishing enterprises after the 1950s, intensive fishing has spread from a few concentrated areas to encompass nearly all fisheries. The scraping of the ocean floor in bottom dragging is devastating to coral, sponges and other long-lived species that do not recover quickly.

This important resource book highlights various processes, particularly sediment transport, coastal erosion and its management, landform evolution and resources of the coastal region.

I thank all members of my team who have helped in the preparation of the book. My special thanks go to "Random Publications" who have published the book.

— S.K. Singh

Contents

Chapter 1

Introduction

A coastline or a seashore is the area where land meets the sea or ocean. A precise line that can be called a coastline cannot be determined due to the Coastline paradox.

Figure: *Rugged coastline of the West Coast Region of New Zealand*

The term "coastal zone" is a region where interaction of the sea and land processes occurs. Both the terms coast and coastal are often used to describe a geographic location or region; for example, New Zealand's West Coast, or the East and West Coasts of the United States.

A pelagic coast refers to a coast which fronts the open ocean, as opposed to a more sheltered coast in a gulf or bay. A shore, on the other hand, can refer to parts of the land which adjoin any large body of water, including oceans (sea shore) and lakes (lake shore). Similarly, the somewhat related term "bank" refers to the land alongside or sloping down to a river (riverbank) or to a body of water smaller than a lake. "Bank" is also used in some parts of the world to refer to an artificial ridge of earth intended to retain the water of a river or pond; in other places this may be called a levee.

While many scientific experts might agree on a common definition of the term "coast", the delineation of the extents of a coast differ according to jurisdiction, with many scientific and government authorities in various countries differing for economic and social policy reasons. According to the UN atlas, 44% of people live within 150 kilometres (93 miles) of the sea.

Formation

Tides often determine the range over which sediment is deposited or eroded. Areas with high tidal ranges allow waves to reach farther up the shore, and areas with lower tidal ranges produce deprossosition at a smaller elevation interval. The tidal range is influenced by the size and shape of the coastline. Tides do not typically cause erosion by themselves; however, tidal bores can erode as the waves surge up river estuaries from the ocean.

Figure: *Atlantic rocky coastline, showing a surf area. Porto Covo, west coast of Portugal*

Waves erode coastline as they break on shore releasing their energy; the larger the wave the more energy it releases and the more sediment it moves. Coastlines with longer shores have more room for the waves to disperse their energy, while coasts with cliffs and short shore faces give little room for the wave energy to be dispersed. In these areas the wave energy breaking against the cliffs is higher, and air and water are compressed into cracks in the rock, forcing the rock apart, breaking it down. Sediment deposited by waves comes from eroded cliff faces and is moved along the coastline by the waves. This forms an abrasion or cliffed coast.

Sediment deposited by rivers is the dominant influence on the amount of sediment located on a coastline. Today riverine deposition at the coast is often blocked by dams and other human regulatory devices, which remove the sediment from the stream by causing it to be deposited inland.

Like the ocean which shapes them, coasts are a dynamic environment with constant change. The Earth's natural processes, particularly sea level rises, waves and various weather phenomena, have resulted in the erosion, accretion and reshaping of coasts as well as flooding and creation of continental shelves and drowned river valleys (rias).

Environmental importance

The coast and its adjacent areas on and off shore are an important part of a local ecosystem: the mixture of fresh water and salt water in estuaries provides many nutrients for marine life. Salt marshes and beaches also support a diversity of plants, animals and insects crucial to the food chain.

Figure: *Somalia has the longest coastline in Africa*

The high level of biodiversity creates a high level of biological activity, which has attracted human activity for thousands of years.

Human Impacts

Human Uses of Coasts

More and more of the world's people live in coastal regions. Many major cities are on or near good harbors and have port facilities. Some landlocked places have achieved port status by building canals.

Figure: *A settled coastline in Marblehead, Massachusetts. Once a fishing port, the harbor is now dedicated to tourism and pleasure boating. Observe that the sand and rocks have been darkened by oil slick up to the high-water line.*

The coast is a frontier that nations have typically defended against military invaders, smugglers and illegal migrants. Fixed coastal defenses have long been erected in many nations and coastal countries typically have a navy and some form of coast guard.

Coasts, especially those with beaches and warm water, attract tourists. In many island nations such as those of the Mediterranean, South Pacific and Caribbean, tourism is central to the economy. Coasts offer recreational activities such as swimming, fishing, surfing, boating, and sunbathing. Growth management can be a challenge for coastal local authorities who often struggle to provide the infrastructure required by new residents.

Threats to a Coast

Coasts also face many human-induced environmental impacts. The human influence on climate change is thought to contribute to an accelerated trend in sea level rise which threatens coastal habitats.

Figure: *This stretch of coast in Tanzania's capital Dar es Salaam serves as a public waste dump.*

Pollution can occur from a number of sources: garbage and industrial debris; the transportation of petroleum in tankers, increasing the probability of large oil spills; small oil spills created by large and small vessels, which flush bilge water into the ocean.

Fishing has declined due to habitat degradation, overfishing, trawling, bycatch and climate change. Since the growth of global fishing enterprises after the 1950s, intensive fishing has spread from a few concentrated areas to encompass nearly all fisheries. The scraping of the ocean floor in bottom dragging is devastating to coral, sponges and other long-lived species that do not recover quickly. This destruction alters the functioning of the ecosystem and can permanently alter species composition and biodiversity. Bycatch, the capture of unintended species in the course of fishing, is typically returned to the ocean only to die from injuries or exposure. Bycatch represents about a quarter of all marine catch. In the case of shrimp capture, the bycatch is five times larger than the shrimp caught.

It is believed that melting Arctic ice will cause sea levels to rise and flood costal areas.

Conservation

Extraordinary population growth in the 20th century has placed stress on the planet's ecosystems. For example, on Saint Lucia, harvesting mangrove for timber and clearing for fishing reduced the mangrove forests, resulting in a loss of habitat and spawning grounds for marine life that was unique to the area. These forests also helped to stabilize the coastline. Conservation efforts since the 1980s have partially restored the ecosystem.

Types of Coast

According to one principle of classification, an emergent coastline is a coastline which has experienced a fall in sea level, because of either a global sea level change, or local uplift. Emergent coastlines are identifiable by the coastal landforms, which are above the high tide mark, such as raised beaches. In contrast, a submergent coastline is one where the sea level has risen, due to a global sea level change, local subsidence, or isostatic rebound. Submergent coastlines are identifiable by their submerged, or "drowned" landforms, such as rias (drowned valleys) and fjords.

According to a second principle of classification, a concordant coastline is a coastline where bands of different rock types run parallel to the shore. These rock types are usually of varying resistance, so the coastline forms distinctive landforms, such as coves. Discordant coastlines feature distinctive landforms because the rocks are eroded by ocean waves. The less resistant rocks erode faster, creating inlets or bays; the more resistant rocks erode more slowly, remaining as headlands or outcroppings.

Other Coastal Categories

- A cliffed coast or abrasion coast is one where marine action has produced steep declivities known as cliffs.
- A flat coast is one where the land gradually descends into the sea.
- A graded shoreline is one where wind and water action has produced a flat and straight coastline.

Coastal Landforms

The following articles describe some coastal landforms

- Bay
- Cape
- Cove
- Gulf
- Headland
- Peninsula

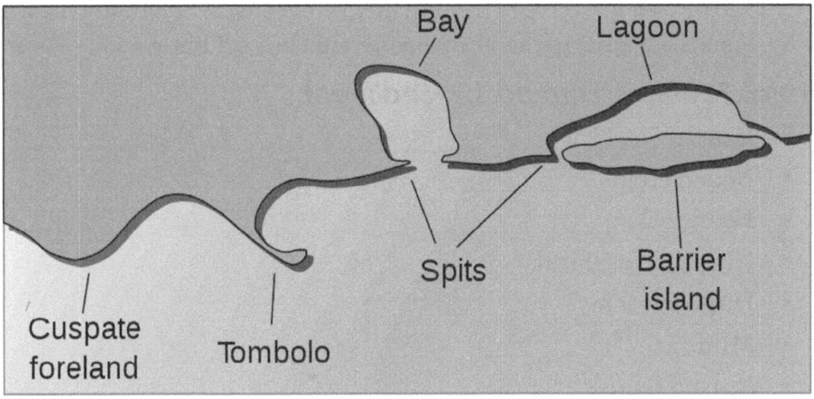

Figure: *Coastal landforms. The feature shown here as a bay would, in certain (mainly southern) parts of Britain, be called a cove. That between the cuspate foreland and the tombolo is a British bay.*

Cliff Erosion

- Much of the sediment deposited along a coast is the result of erosion of a surrounding cliff, or bluff. Sea cliffs retreat landward because of the constant undercutting of slopes by waves. If the slope/cliff being undercut is made of unconsolidated sediment it will erode at a much faster rate then a cliff made of bedrock.
- A natural arch is formed when a sea stacks is eroded through by waves.
- Sea caves are made when certain rock beds are more susceptible to erosion than the surrounding rock beds because of different areas of weakness. These areas are eroded at a faster pace creating a hole or crevasse that, through time, by means of wave action and erosion, becomes a cave.
- A stack is formed when a headland is eroded away by wave and wind action.
- A stump is a shortened sea stack that has been eroded away or fallen because of instability.
- Wave-cut notches are caused by the undercutting of overhanging slopes which leads to increased stress on cliff material and a greater probability that the slope material will fall. The fallen debris accumulates at the bottom of the cliff and is eventually removed by waves.
- A wave-cut platform forms after erosion and retreat of a sea cliff has been occurring for a long time. Gently sloping wave-cut platforms develop early on in the first stages of cliff retreat.

Later the length of the platform decreases because the waves lose their energy as they break further off shore.

Coastal Features Formed by Sediment

- Beach
- Beach cusps
- Boondocks
- Cuspate foreland
- Dune system
- Mudflat
- Raised beach
- Ria
- Shoal
- Strand plain
- Spit
- Surge channel
- Tombolo

Coastal Features Formed by Another Feature

- Lagoon
- Salt marsh

Other Features on the Coast

- Concordant coastline
- Discordant coastline
- Fjord
- Island
- Island arc
- Machair

Coastal Processes

The following articles describe the various geologic processes that affect a coastal zone:

- Attrition
- Currents
- Denudation
- Deposition
- Erosion

- Flooding
- Longshore drift
- Saltation
- Sea level change
    - o eustatic
    - o isostatic
- Sedimentation
- Coastal sediment supply
    - o sediment transport
    - o solution
    - o subaerial processes
    - o suspension
- Tides
- Water waves
    - o diffraction
    - o refraction
    - o wave breaking
    - o wave shoaling
- Weathering

Wildlife

Animals

Some of the animals live along a typical coast. There are animals like puffins, sea turtles and rockhopper penguins. Sea snails and various kinds of barnacles live on the coast and scavenge on food deposited by the sea. Most coastal animals are used to humans in developed areas, such as dolphins and seagulls who eat food thrown for them by tourists. Since the coastal areas are all part of the littoral zone, there is a profusion of marine life found just off-coast.

There are many kinds of seabirds on the coast. Pelicans and cormorants join up with terns and oystercatchers to forage for fish and shellfish on the coast. There are sea lions on the coast of Wales and other countries.

Plants

Coastal areas are famous for their kelp beds. Kelp is a fast-growing seaweed that grows up to a metre a day. Corals and sea anemones are true animals, but live a lifestyle similar to that of plants. Mangroves, seagrasses and salt marsh are important coastal vegetation types in tropical and temperate environments respectively.

Coastline Statistics

Coastline Problem

Shortly before 1951, Lewis Fry Richardson, in researching the possible effect of border lengths on the probability of war, noticed that the Portuguese reported their measured border with Spain to be 987 km, but the Spanish reported it as 1214 km. This was the beginning of the coastline problem, which is a mathematical uncertainty inherent in the measurement of boundaries that are irregular.

The prevailing method of estimating the length of a border (or coastline) was to lay out *n* equal straight-line segments of length –*!* with dividers on a map or aerial photograph. Each end of the segment must be on the boundary. Investigating the discrepancies in border estimation, Richardson discovered what is now termed the Richardson Effect: the sum of the segments is inversely proportional to the common length of the segments. In effect, the shorter the ruler, the longer the measured border; the Spanish and Portuguese geographers were simply using different-length rulers.

The result most astounding to Richardson is that, under certain circumstances, as –*!* approaches zero, the length of the coastline approaches infinity. Richardson had believed, based on Euclidean geometry, that a coastline would approach a fixed length, as do similar estimations of regular geometric figures. For example, the perimeter of a regular polygon inscribed in a circle approaches the circumference with increasing numbers of sides (and decrease in the length of one side). In geometric measure theory such a smooth curve as the circle that can be approximated by small straight segments with a definite limit is termed a rectifiable curve.

Measuring a Coastline

More than a decade after Richardson completed his work, Benoit Mandelbrot developed a new branch of mathematics, fractal geometry, to describe just such non-rectifiable complexes in nature as the infinite coastline. His own definition of the new figure serving as the basis for his study is:

I coined *fractal* from the Latin adjective *fractus*. The corresponding Latin verb *frangere* means "to break:" to create irregular fragments. It is therefore sensible ... that, in addition to "fragmented" ... *fractus* should also mean "irregular."

A key property of the fractal is self-similarity; that is, at any scale the same general configuration appears. A coastline is perceived as bays alternating with promontories. In the hypothetical situation that a given coastline has this property of self-similarity, then no matter how greatly any one small section of coastline is magnified, a similar pattern of smaller bays and promontories superimposed on larger bays and promontories appears, right down to the grains of sand. At that scale the coastline appears as a momentarily shifting, potentially infinitely long thread with a stochastic arrangement of bays and promontories formed from the small objects at hand. In such an environment (as opposed to smooth curves) Mandelbrot asserts "coastline length turns out to be an elusive notion that slips between the fingers of those who want to grasp it."

There are different kinds of fractals. A coastline with the stated property is in "a first category of fractals, namely curves whose fractal dimension is greater than 1." That last statement represents an extension by Mandelbrot of Richardson's thought. Mandelbrot's statement of the Richardson Effect is:

$$L(\epsilon) \sim F\epsilon^{1-D}$$

where L, coastline length, a function of the measurement unit, ε, is approximated by the expression. F is a constant and D is a parameter that Richardson found depended on the coastline approximated by L. He gave no theoretical explanation but Mandelbrot identified D with a non-integer form of the Hausdorff dimension, later the fractal dimension. Rearranging the right side of the expression obtains:

$$\frac{F}{\epsilon^{D}} \cdot \epsilon$$

where $F\varepsilon^{-D}$ must be the number of units ε required to obtain L. The fractal dimension is the number of the dimensions of the figure being used to approximate the fractal: 0 for a dot, 1 for a line, 2 for a square. D in the expression is between 1 and 2, for coastlines typically less than 1.5. The broken line measuring the coast does not extend in one direction nor does it represent an area, but is intermediate. It can be interpreted as a thick line or band of width 2ε. More broken coastlines have greater D and therefore L is longer for the same ε. Mandelbrot showed that D is independent of ε.

A coastal plain is an area of flat, low-lying land adjacent to a seacoast. One of the world's largest coastal plains is located in eastern

South America. The Gulf Coastal Plain of North America extends northwards from the Gulf of Mexico along the Lower Mississippi River to the Ohio River, which is a distance of about 500 miles (800 km). During the Cretaceous period, the central area of the United States was covered by a shallow sea, which disappeared as the land rose. Large fossilized aquatic birds called *Hesperornis* and *Ichthyornis*, found in western Kansas, indicate that the shallow sea was rife with fish.

Coastal regions offer some of the most spectacular and dynamic environments to be found anywhere in the world, with 40% of the world's population living within 100km of the coast. They boast stunning cliffs and geological features, beautiful biodiversity, in the form of mangrove forests and coral reefs, including countless World Heritage sites and they are amongst the most rapidly changing environments in the world; whether it be due to human development like that of Dubai or coastal retreat such as that of the Sundarbans, in West Bengal. Climate change threatens coastlines across the globe with sea level rise, rates of erosion, land loss and slianisation threaten the means of millions to support themselves. Coastal environments offer the geographer a stunning choice of places to discover the world and study the interaction between Physical and Human Geography.

Waves and Wavedominated Coasts

In fluid dynamics, wind waves, or wind-generated waves, are surface waves that occur on the free surface of oceans, seas, lakes, rivers, and canals or even on small puddles and ponds. They result from the wind blowing over an area of fluid surface. Waves in the oceans can travel thousands of miles before reaching land. Wind waves range in size from small ripples, to waves over 100ft (30m) high.

When directly generated and affected by local winds, a wind wave system is called a wind sea. After the wind ceases to blow, wind waves are called *swells*. More generally, a swell consists of wind-generated waves that are not significantly affected by the local wind at that time. They have been generated elsewhere or some time ago. Wind waves in the ocean are called ocean surface waves.

Wind waves have a certain amount of randomness: subsequent waves differ in height, duration, and shape with limited predictability. They can be described as a stochastic process, in combination with the physics governing their generation, growth, propagation and decay—as well as governing the interdependence between flow quantities such as: the water surface movements, flow velocities and water pressure.

The key statistics of wind waves (both seas and swells) in evolving sea states can be predicted with wind wave models.

Figure: *Ocean waves*

Although waves are usually considered in the water seas of Earth, the hydrocarbon seas of Titan may also have wind-driven waves.

Wave Formation

The great majority of large breakers one observes on a beach result from distant winds. Five factors influence the formation of wind waves:

- Wind speed or strength relative to wave speed- the wind must be moving faster than the wave crest for energy transfer
- The uninterrupted distance of open water over which the wind blows without significant change in direction (called the *fetch*)
- Width of area affected by fetch
- Wind duration - the time over which the wind has blown over a given area
- Water depth

All of these factors work together to determine the size of wind waves:

- Wave height (from high trough to crest)
- Wave length (from crest to crest)
- Wave period (time interval between arrival of consecutive crests at a stationary point)
- Wave propagation direction

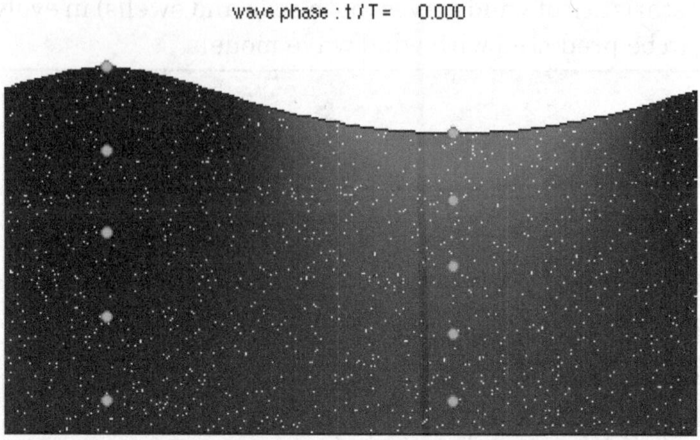

***Figure:** water particles movements of deep a water wave.*

A fully developed sea has the maximum wave size theoretically possible for a wind of a specific strength, duration, and fetch. Further exposure to that specific wind could only cause a loss of energy due to the breaking of wave tops and formation of "whitecaps". Waves in a given area typically have a range of heights. For weather reporting and for scientific analysis of wind wave statistics, their characteristic height over a period of time is usually expressed as *significant wave height*. This figure represents an average height of the highest one-third of the waves in a given time period (usually chosen somewhere in the range from 20 minutes to twelve hours), or in a specific wave or storm system. The significant wave height is also the value a "trained observer" (e.g. from a ship's crew) would estimate from visual observation of a sea state. Given the variability of wave height, the largest individual waves are likely to be somewhat less than twice the reported significant wave height for a particular day or storm.

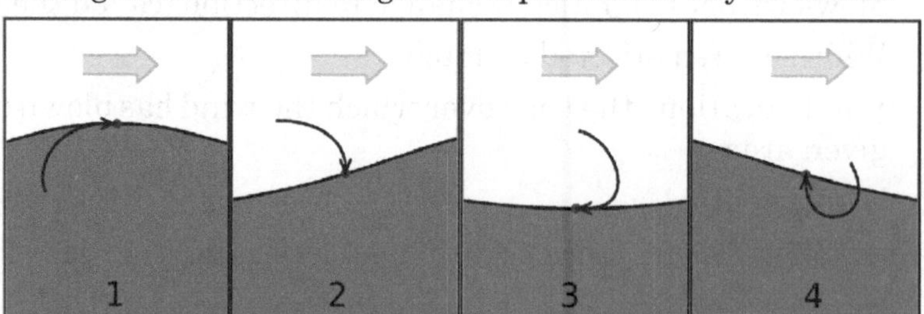

***Figure:** The phases of an ocean surface wave: 1. Wave Crest, where the water masses of the surface layer are moving horizontally in the same direction as the propagating wave front. 2. Falling wave. 3. Trough, where the water masses of the surface layer are moving horizontally in the opposite direction of the wave front direction. 4. Rising wave.*

Sources of wind wave generation: Sea water wave is generated by many kinds of disturbances such as Seismic events, gravity, and crossing wind. The generation of wind wave is initiated by the disturbances of cross wind field on the surface of the sea water. Two major Mechanisms of surface wave formation by winds (a.k.a.'The Miles-Phillips Mechanism') and other sources (ex. earthquakes) of wave formation can explain the generation of wind waves.

However, if one set a flat water surface (Beaufort Point,0) and abrupt cross wind flows on the surface of the water, then the generation of surface wind waves can be explained by following two mechanisms which initiated by normal pressure fluctuations of turbulent winds and parallel wind shear flows.

The Mechanism of the Surface Wave Generation by Winds

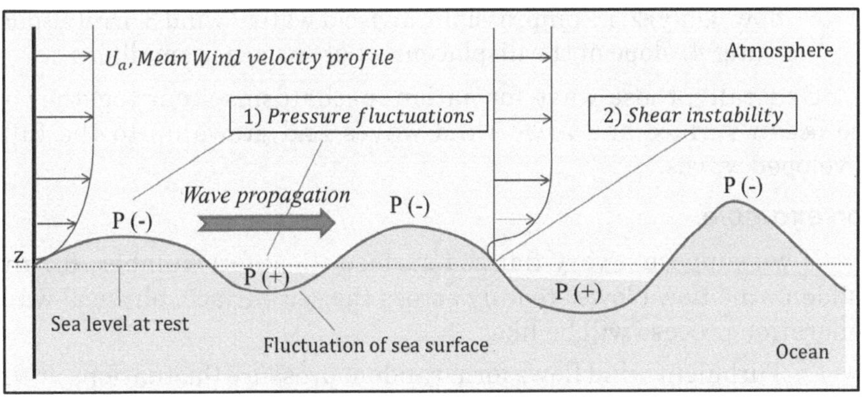

Figure: *the simple picture of the wave formation mechanism*

1) Starts from "Fluctuations of wind" (O.M.Phillips) : the wind wave formation on water surface by wind is started by a random distribution of normal pressure acting on the water from the wind. By the mechanism developed by O.M. Phillips (in 1957), the water surface is initially at rest and the generation of wave is initiated by adding turbulent wind flows and then, by the fluctuations of the wind, normal pressure acting on the water surface. This pressure fluctuation arise normal and tangential stresses to the surface water, and generates wave behavior on the water surface. {*Assumptions* 1. water originally at rest 2. water is inviscid 3. Water is irrotational 4. Random distribution of normal pressure to the water surface from the turbulent wind 5. Correlations between air and Water motions are neglected}
2) starts from "wind shear forces" on the water surface (J.W.Miles, applied to mainly 2D deep water gravity waves) ; John W. Miles

suggested a surface wave generation mechanism which is initiated by turbulent wind shear flows Ua(y), based on the inviscid Orr-Sommerfeld equation in 1957. He found the energy transfer from wind to water surface as a wave speed, c is proportional to the curvature of the velocity profile of wind Ua"(y) at point where the mean wind speed is equal to the wave speed (Ua=c, where, Ua is the Mean turbulent wind speed). Since the wind profile Ua(y) is logarithmic to the water surface, the curvature Ua"(y) have negative sign at the point of Ua=c. This relations show the wind flow transferring its kinetic energy to the water surface at their interface, and arises wave speed, c. the growth-rate can be determined by the curvature of the winds ((d^2 Ua)/(dz^2)) at the steering height (Ua (z=z_h)=c) for a given wind speed Ua {*Assumptions*; 1. 2D parallel shear flow, Ua(y) 2. incompressible, inviscid water / wind 3. irrotational water 4. slope of the displacement of surface is small}

Generally, these wave formation mechanisms occur together on the ocean surface and arise wind waves and grows up to the fully developed waves.

For example,

If we suppose a very flat sea surface (Beaufort number, 0), and sudden wind flow blows steadily across the sea surface, physical wave generation process will be like;

1. Turbulent wind flows form random pressure fluctuations at the sea surface. Small waves with a few centimeters order of wavelengths is generated by the pressure fluctuations. (The Phillips mechanism)
2. The cross wind keep acting on the initially fluctuated sea surface, then the wave become larger. As the wave become larger, the pressure differences get larger along to the wave growing, then the wave growth rate is getting faster. Then the shear instability expedites the wave growing exponentially. (The Miles mechanism)
3. The interactions between the waves on the surface generate longer waves (Hasselmann et al., 1973) and the interaction will transfer wave energy from the shorter waves generated by the Miles mechanism to the waves have slightly lower frequencies than the frequency at the peak wave magnitudes, then finally the waves will be faster than the cross wind speed (Pierson & Moskowitz).

Conditions Necessary for a Fully Developed Sea at Given Wind Speeds, and the Parameters of the Resulting Waves					
Wind Conditions			***Wave Size***		
Wind Speed in One Direction	Fetch	Wind Duration	Average Height	Average Wavelength	Average Period and Speed
19 km/h (12 mph)	19 km (12 mi)	2 hr	0.27 m (0.9 ft)	8.5 m (28 ft)	3.0 sec 9.3 ft/sec
37 km/h (23 mph)	139 km (86 mi)	10 hr	1.5 m (4.9 ft)	33.8 m (111 ft)	5.7 sec 19.5 ft/sec
56 km/h (35 mph)	518 km (322 mi)	23 hr	4.1 m (13.6 ft)	76.5 m (251 ft)	8.6 sec 29.2 ft/sec
74 km/h (46 mph)	1,313 km (816 mi)	42 hr	8.5 m (27.9 ft)	136 m (446 ft)	11.4 sec 39.1 ft/sec
92 km/h (58 mph)	2,627 km (1,633 mi)	69 hr	14.8 m (48.7 ft)	212.2 m (696 ft)	14.3 sec 48.7 ft/sec

((NOTE: Most of the wave speeds calculated from the wave length divided by the period are proportional to sqrt (length).

Types of Wind Waves

Three different types of wind waves develop over time:

- Capillary waves, or ripples
- Seas
- Swells

Ripples appear on smooth water when the wind blows, but will die quickly if the wind stops. The restoring force that allows them to propagate is surface tension. Seas are the larger-scale, often irregular motions that form under sustained winds. These waves tend to last much longer, even after the wind has died, and the restoring force that allows them to propagate is gravity. As waves propagate away from their area of origin, they naturally separate into groups of common direction and wavelength. The sets of waves formed in this way are known as swells.

Individual "rogue waves" (also called "freak waves", "monster waves", "killer waves", and "king waves") much higher than the other waves in the sea state can occur. In the case of the Draupner wave, its 25 m (82 ft) height was 2.2 times the significant wave height. Such waves are distinct from tides, caused by the Moon and Sun's gravitational pull, tsunamis that are caused by underwater earthquakes or landslides, and waves generated by underwater explosions or the fall of meteorites—all having far longer wavelengths than wind waves.

Yet, the largest ever recorded wind waves are common — not rogue — waves in extreme sea states. For example: 29.1 m (95 ft) high waves have been recorded on the RRS Discovery in a sea with 18.5 m (61 ft) significant wave height, so the highest wave is only 1.6 times the significant wave height. The biggest recorded by a buoy (as of 2011) was 32.3 m (106 ft) high during the 2007 typhoon Krosa near Taiwan.

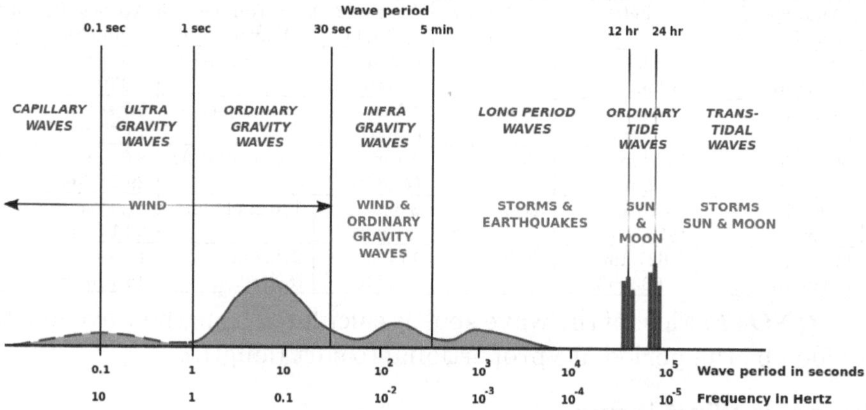

Figure: *Classification of the spectrum of ocean waves according to wave period.*

Ocean waves can be classified based on: the disturbing force(s) that create(s) them; the extent to which the disturbing force(s) continue(s) to influence them after formation; the extent to which the restoring force(s) weaken(s) (or flatten) them; and their wavelength or period. Seismic Sea waves have a period of ~20 minutes, and speeds of 760km/h (470mph). Wind waves (deep-water waves) have a period of about 20 seconds.

Wave type	*Typical wavelength*	*Disturbing force*	*Restoring force*
Capillary wave	< 2 cm	Wind	Surface tension
Wind wave	60–150 m (200–500 ft)	Wind over ocean	Gravity
Seiche	Large, variable; a function of basin size	Change in atmospheric pressure, storm surge	Gravity
Seismic sea wave (tsunami)	200 km (125 mi)	Faulting of sea floor, volcanic eruption, landslide	Gravity
Tide	Half the circumference of Earth	Gravitational attraction, rotation of Earth	Gravity

The speed of all ocean waves is controlled by gravity, wavelength, and water depth. Most characteristics of ocean waves depend on the relationship between their wavelength and water depth. Wavelength determines the size of the orbits of water molecules within a wave, but water depth determines the shape of the orbits. The paths of water molecules in a wind wave are circular only when the wave is traveling in deep water. A wave cannot "feel" the bottom when it moves through water deeper than half its wavelength because too little wave energy is contained in the small circles below that depth. Waves moving through water deeper than half their wavelength are known as deep-water waves. On the other hand, the orbits of water molecules in waves moving through shallow water are flattened by the proximity of the sea surface bottom. Waves in water shallower than 1/20 their original wavelength are known as shallow-water waves. Transitional waves travel through water deeper than 1/20 their original wavelength but shallower than half their original wavelength.

In general, the longer the wavelength, the faster the wave energy will move through the water. For deep-water waves, this relationship is represented with the following formula:

$$C = L / T$$

where C is speed (celerity), L is wavelength, and T is time, or period (in seconds).

The speed of a deep-water wave may also be approximated by:

$$C = \sqrt{gL / 2\pi}$$

where g is the acceleration due to gravity, 9.8 meters (32.2 feet) per second squared. Because g and π (3.14) are constants, the equation can be reduced to:

$$C = 1.251\sqrt{L}$$

when C is measured in meters per second and L in meters. Note that in both instances that wave speed is proportional to wavelength.

The speed of shallow-water waves is described by a different equation that may be written as:

$$C = \sqrt{gd} = 3.1\sqrt{d}$$

where C is speed (in meters per second), g is the acceleration due to gravity, and d is the depth of the water (in meters). The period of a wave remains unchanged regardless of the depth of water through which it is moving. As deep-water waves enter the shallows and feel the bottom, however, their speed is reduced and their crests "bunch up," so their wavelength shortens.

Wave Shoaling and Refraction

As waves travel from deep to shallow water, their shape alters (wave height increases, speed decreases, and length decreases as wave orbits become asymmetrical). This process is called shoaling.

Wave refraction is the process by which wave crests realign themselves as a result of decreasing water depths. Varying depths along a wave crest cause the crest to travel at different phase speeds, with those parts of the wave in deeper water moving faster than those in shallow water. This process continues until the crests become (nearly) parallel to the depth contours. Rays—lines normal to wave crests between which a fixed amount of energy flux is contained—converge on local shallows and shoals. Therefore, the wave energy between rays is concentrated as they converge, with a resulting increase in wave height.

Because these effects are related to a spatial variation in the phase speed, and because the phase speed also changes with the ambient current – due to the Doppler shift – the same effects of refraction and altering wave height also occur due to current variations. In the case of meeting an adverse current the wave *steepens*, i.e. its wave height increases while the wave length decreases, similar to the shoaling when the water depth decreases.

Wave Breaking

Some waves undergo a phenomenon called "breaking". A breaking wave is one whose base can no longer support its top, causing it to collapse. A wave breaks when it runs into shallow water, or when two wave systems oppose and combine forces. When the slope, or steepness ratio, of a wave is too great, breaking is inevitable.

Individual waves in deep water break when the wave steepness—the ratio of the wave height H to the wavelength λ—exceeds about 0.07, so for $H > 0.07\ \lambda$. In shallow water, with the water depth small compared to the wavelength, the individual waves break when their wave height H is larger than 0.8 times the water depth h, that is

$H > 0.8\ h$. Waves can also break if the wind grows strong enough to blow the crest off the base of the wave.

Three main types of breaking waves are identified by surfers or surf lifesavers. Their varying characteristics make them more or less suitable for surfing, and present different dangers.

- Spilling, or rolling: these are the safest waves on which to surf. They can be found in most areas with relatively flat shorelines. They are the most common type of shorebreak
- Plunging, or dumping: these break suddenly and can "dump" swimmers—pushing them to the bottom with great force. These are the preferred waves for experienced surfers. Strong offshore winds and long wave periods can cause dumpers. They are often found where there is a sudden rise in the sea floor, such as a reef or sandbar.
- Surging: these may never actually break as they approach the water's edge, as the water below them is very deep. They tend to form on steep shorelines. These waves can knock swimmers over and drag them back into deeper water.

Science of Waves

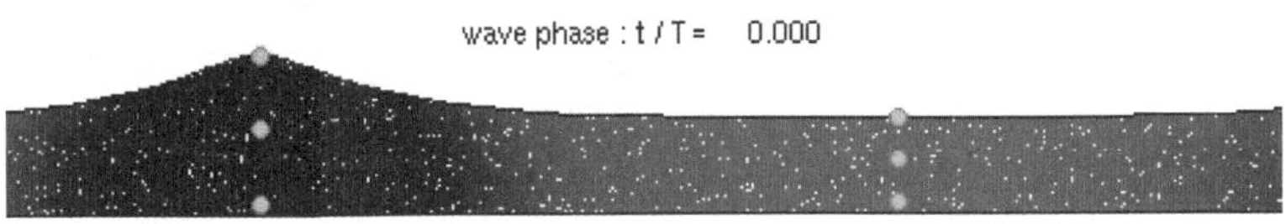

Figure: *Stokes drift in shallow water waves (Animation)*

Wind waves are mechanical waves that propagate along the interface between water and air; the restoring force is provided by gravity, and so they are often referred to as surface gravity waves. As the wind blows, pressure and friction forces perturb the equilibrium of the water surface. These forces transfer energy from the air to the water, forming waves. The initial formation of waves by the wind is described in the theory of Phillips from 1957, and the subsequent growth of the small waves has been modeled by Miles, also in 1957.

In the case of monochromatic linear plane waves in deep water, particles near the surface move in circular paths, making wind waves a combination of longitudinal (back and forth) and transverse (up and down) wave motions. When waves propagate in shallow water, (where the depth is less than half the wavelength) the particle trajectories are compressed into ellipses.

As the wave amplitude (height) increases, the particle paths no longer form closed orbits; rather, after the passage of each crest, particles are displaced slightly from their previous positions, a phenomenon known as Stokes drift.

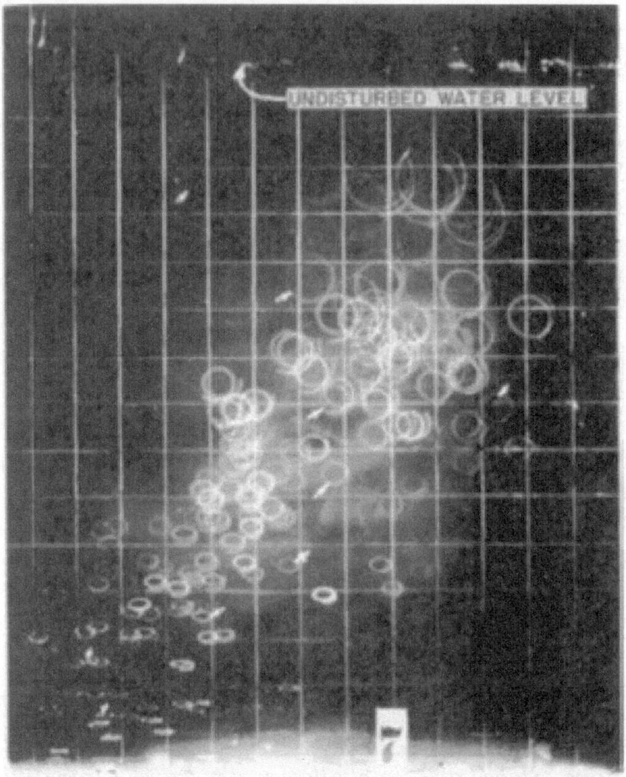

***Figure:** Photograph of the water particle orbits under a – progressive and periodic – surface gravity wave in a wave flume. The wave conditions are: mean water depth d = 2.50 ft (0.76 m), wave height H = 0.339 ft (0.103 m), wavelength λ = 6.42 ft (1.96 m), period T = 1.12 s.*

As the depth below the free surface increases, the radius of the circular motion decreases. At a depth equal to half the wavelength λ, the orbital movement has decayed to less than 5% of its value at the surface. The phase speed (also called the celerity) of a surface gravity wave is – for pure periodic wave motion of small-amplitude waves – well approximated by

$$c = \sqrt{\frac{g\lambda}{2\pi}\tanh\left(\frac{2\pi d}{\lambda}\right)}$$

where

c = phase speed;

λ = wavelength;

d = water depth;

g = acceleration due to gravity at the Earth's surface.

In deep water, where $d \geq \frac{1}{2}\lambda$, so $\frac{2\pi d}{\lambda} \geq \pi$ and the hyperbolic tangent approaches 1, the speed c approximates

$$c_{\text{deep}} = \sqrt{\frac{g\lambda}{2\pi}}.$$

In SI units, with c_{deep} in m/s, $c_{\text{deep}} \approx 1.25\sqrt{\lambda}$, when λ is measured in metres. This expression tells us that waves of different wavelengths travel at different speeds. The fastest waves in a storm are the ones with the longest wavelength. As a result, after a storm, the first waves to arrive on the coast are the long-wavelength swells.

For intermediate and shallow water, the Boussinesq equations are applicable, combining frequency dispersion and nonlinear effects. And in very shallow water, the shallow water equations can be used.

If the wavelength is very long compared to the water depth, the phase speed (by taking the limit of c when the wavelength approaches infinity) can be approximated by

$$c_{\text{shallow}} = \lim_{\lambda \to \infty} c = \sqrt{gd}.$$

On the other hand, for very short wavelengths, surface tension plays an important role and the phase speed of these gravity-capillary waves can (in deep water) be approximated by

$$c_{\text{gravity-capillary}} = \sqrt{\frac{g\lambda}{2\pi} + \frac{2\pi S}{\rho\lambda}}$$

where

S = surface tension of the air-water interface;

ρ = density of the water.

When several wave trains are present, as is always the case in nature, the waves form groups. In deep water the groups travel at a group velocity which is half of the phase speed. Following a single wave in a group one can see the wave appearing at the back of the group, growing and finally disappearing at the front of the group.

As the water depth d decreases towards the coast, this will have an effect: wave height changes due to wave shoaling and refraction. As the wave height increases, the wave may become unstable when the crest of the wave moves faster than the trough. This causes *surf*, a breaking of the waves. The movement of wind waves can be captured by wave energy devices. The energy density (per unit area) of regular sinusoidal waves depends on the water density ρ, gravity acceleration g and the wave height H (which, for regular waves, is equal to twice the amplitude, a):

$$E = \frac{1}{8}\rho g H^2 = \frac{1}{2}\rho g a^2.$$

The velocity of propagation of this energy is the group velocity.

Breaking Wave

In fluid dynamics, a breaking wave is a wave whose amplitude reaches a critical level at which some process can suddenly start to occur that causes large amounts of wave energy to be transformed into turbulent kinetic energy. At this point, simple physical models that describe wave dynamics often become invalid, particularly those that assume linear behaviour.

***Figure:** A large wave breaking.*

The most generally familiar sort of breaking wave is the breaking of water surface waves on a coastline. Because of the horizontal component of the fluid velocity associated with the wave motion, wave crests steepen as the amplitude increases; wave breaking generally occurs where the amplitude reaches the point that the crest of the wave actually overturns—though the types of breaking water surface waves are discussed in more detail below. Certain other effects in fluid dynamics have also been termed "breaking waves," partly by analogy with water surface waves. In meteorology, atmospheric gravity waves are said to break when the wave produces regions where the potential temperature decreases with height, leading to energy dissipation through convective instability; likewise Rossby waves are said to break when the potential vorticity gradient is overturned. Wave breaking also occurs in plasmas, when the particle velocities exceed the wave's phase speed.

Figure: *Breaking waves at the village of Porto Covo, west coast of Portugal.*

Breaking Water Surface Waves

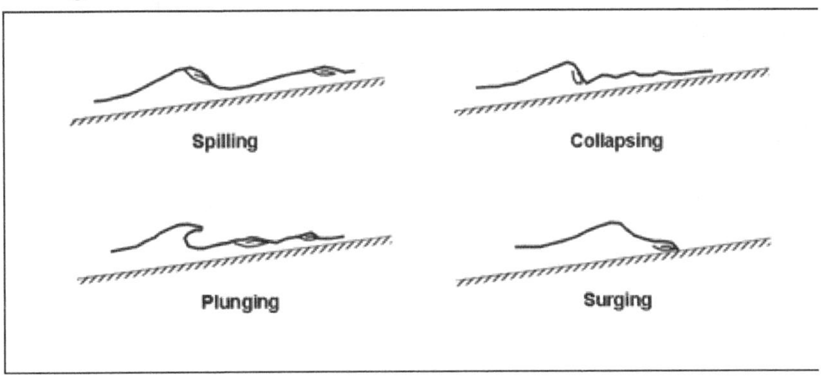

Figure: *Classification of breaking wave types.*

Breaking of water surface waves may occur anywhere that the amplitude is sufficient, including in mid-ocean. However, it is particularly common on beaches because wave heights are amplified in the region of shallower water (because the group velocity is lower there). There are four basic types of breaking water waves. They are spilling, plunging, collapsing, and surging.

Spilling Breakers

When the ocean floor has a gradual slope, the wave will steepen until the crest becomes unstable, resulting in turbulent whitewater spilling down the face of the wave. This continues as the wave approaches the shore, and the wave's energy is slowly dissipated in the whitewater. Because of this, spilling waves break for a longer time than other waves, and create a relatively gentle wave. Onshore wind conditions make spillers more likely.

Plunging Breakers

A plunging wave occurs when the ocean floor is steep or has sudden depth changes, such as from a reef or sandbar. The crest of the wave becomes much steeper than a spilling wave, becomes vertical, then curls over and drops onto the trough of the wave, releasing most of its energy at once in a relatively violent impact. A plunging wave breaks with more energy than a significantly larger spilling wave. The wave can trap and compress the air under the lip, which creates the "crashing" sound associated with waves. With large waves, this crash can be felt by beachgoers on land. Offshore wind conditions can make plungers more likely.

If a plunging wave is not parallel to the beach (or the ocean floor), the section of the wave which reaches shallow water will break first, and the breaking section (or curl) will move laterally across the face of the wave as the wave continues. This is the "tube" that is so highly sought after by surfers (also called a "barrel", a "pit", and "the greenroom", among other terms). The surfer tries to stay near or under the curling lip, often trying to stay as "deep" in the tube as possible while still being able to shoot forward and exit the tube before it closes. A plunging wave that is parallel to the beach can break along its whole length at once, rendering it unrideable and dangerous. Surfers refer to these waves as "closed out".

Collapsing

Collapsing waves are a cross between plunging and surging, in which the crest never fully breaks, yet the bottom face of the wave gets steeper and collapses, resulting in foam.

Surging

Surging breakers originate from long period, low steepness waves and/or steep beach profiles. The outcome is the rapid movement of the base of the wave up the swash slope and the disappearance of the wave crest. The front face and crest of the wave remain relatively smooth with little foam or bubbles, resulting in a very narrow surf zone, or no breaking waves at all.

The short, sharp burst of wave energy means that the swash/ backwash cycle completes before the arrival of the next wave, leading to a low value of Kemp's phase difference (< 0.5). Surging waves are typical of reflective beach states. On steeper beaches, the energy of the wave can be reflected by the bottom back into the ocean, causing standing waves.

Physics

During breaking, a deformation (usually a bulge) forms at the wave crest, either leading side of which is known as the "toe." Parasitic capillary waves are formed, with short wavelengths. Those above the "toe" tend to have much longer wavelengths. This theory is anything but perfect, however, as it's linear. There have been a couple non-linear theories of motion (regarding waves). One put forth uses a perturbation method to expand the description all the way to the third order, and better solutions have been found since then. As for wave deformation, methods much like the boundary integral method and the Boussinesq model have been created.

It has been found that high-frequency detail present in a breaking wave plays a part in crest deformation and destabilization. The same theory expands on this, stating that the valleys of the capillary waves create a source for vorticity. It is said that surface tension (and viscosity) are significant for waves up to 2m in wavelength.

These models are flawed, however, as they can't take into account what happens to the water after the wave breaks. Post-break eddy forms and the turbulence created via the breaking is mostly unresearched. Understandably, it might be difficult to glean predictable results from the ocean.

After the tip of the wave overturns and the jet collapses, it creates a very coherent and defined horizontal vertex. The plunging breakers create secondary eddies down the face of the wave. Small horizontal random eddides that form on the sides of the wave suggest that, perhaps, prior to breaking, the water's velocity is more or less two dimensional. This becomes three dimensional upon breaking.

The main vortex along the front of the wave diffuses rapidly into the interior of the wave after breaking, as the eddies on the surface become more viscous. Advection and molecular diffusion play a part in stretching the vortex and redistributing the vorticity, as well as the formation turbulence cascades. The energy of the large vortices are, by this method, is transferred to much smaller isotropic vortices.

Experiments have been conducted to deduce the evolution of turbulence after break, both in deep water and on a beach.

Wind Wave Models

Surfers are very interested in the wave forecasts. There are many websites that provide predictions of the surf quality for the upcoming days and weeks. Wind wave models are driven by more general weather models that predict the winds and pressures over the oceans, seas and lakes.

Wind wave models are also an important part of examining the impact of shore protection and beach nourishment proposals. For many beach areas there is only patchy information about the wave climate, therefore estimating the effect of wind waves is important for managing littoral environments.

Seismic Signals

Ocean water waves generate land seismic waves that propagate hundreds of kilometers into the land. These seismic signals usually have the period of 6 ± 2 seconds. Such recordings were first reported and understood in about 1900.

There are two types of seismic "ocean waves". The primary waves are generated in shallow waters by direct water wave-land interaction and have the same period as the water waves (10 to 16 seconds). The more powerful secondary waves are generated by the superposition of ocean waves of equal period traveling in opposite directions, thus generating standing gravity waves – with an associated pressure oscillation at half the period, which is not diminishing with depth. The theory for microseism generation by standing waves was provided by Michael Longuet-Higgins in 1950, after in 1941 Pierre Bernard suggested this relation with standing waves on the basis of observations.

Internal Waves

Internal waves can form at the boundary between water layers of different densities. These sub-surface waves are called internal waves. As is the case with ocean waves at the air-ocean interface, internal waves possess troughs, crests, wavelength, and period. Internal waves move very slowly because the density difference between the joined media is very small. Internal waves occur in the ocean at the base of the pycnocline, especially at the bottom edge of a steep thermocline. The wave height of internal waves may be greater than 30 meters (100 feet), causing the pycnocline to undulate slowly through a considerable depth. Their wavelength often exceeds 0.8 kilometres (0.50 mi) and their periods are typically 5 to 8 minutes. Internal waves are generated by wind energy, tidal energy, and ocean currents. Surface manifestations of internal waves have been photographed from space.

Internal waves may mix nutrients into surface water and trigger plankton blooms. They can also affect submarines and oil platforms.

Chapter 2

Marine Pollution

Marine pollution occurs when harmful, or potentially harmful, effects result from the entry into the ocean of chemicals, particles, industrial, agricultural and residential waste, noise, or the spread of invasive organisms.

Figure: *While marine pollution can be obvious, as with the marine debris shown above, it is often the pollutants that cannot be seen that cause most harm.*

Most sources of marine pollution are land based. The pollution often comes from nonpoint sources such as agricultural runoff, wind-blown debris and dust. Nutrient pollution, a form of water pollution, refers to contamination by excessive inputs of nutrients. It is a primary cause of eutrophication of surface waters, in which excess nutrients, usually nitrogen or phosphorus, stimulate algae growth.

Many potentially toxic chemicals adhere to tiny particles which are then taken up by plankton and benthos animals, most of which are either deposit or filter feeders. In this way, the toxins are concentrated upward within ocean food chains. Many particles combine chemically in a manner highly depletive of oxygen, causing estuaries to become anoxic.

When pesticides are incorporated into the marine ecosystem, they quickly become absorbed into marine food webs. Once in the food webs, these pesticides can cause mutations, as well as diseases, which can be harmful to humans as well as the entire food web.

Toxic metals can also be introduced into marine food webs. These can cause a change to tissue matter, biochemistry, behaviour, reproduction, and suppress growth in marine life. Also, many animal feeds have a high fish meal or fish hydrolysate content. In this way, marine toxins can be transferred to land animals, and appear later in meat and dairy products.

History

Although marine pollution has a long history, significant international laws to counter it were only enacted in the twentieth century. Marine pollution was a concern during several United Nations Conferences on the Law of the Sea beginning in the 1950s. Most scientists believed that the oceans were so vast that they had unlimited ability to dilute, and thus render pollution harmless.

In the late 1950s and early 1960s, there were several controversies about dumping radioactive waste off the coasts of the United States by companies licensed by the Atomic Energy Commission, into the Irish Sea from the British reprocessing facility at Windscale, and into the Mediterranean Sea by the French Commissariat à l'Energie Atomique. After the Mediterranean Sea controversy, for example, Jacques Cousteau became a worldwide figure in the campaign to stop marine pollution. Marine pollution made further international headlines after the 1967 crash of the oil tanker Torrey Canyon, and after the 1969 Santa Barbara oil spill off the coast of California.

Marine pollution was a major area of discussion during the 1972 United Nations Conference on the Human Environment, held in Stockholm. That year also saw the signing of the Convention on the Prevention of Marine Pollution by Dumping of Wastes and Other Matter, sometimes called the London Convention. The London Convention did not ban marine pollution, but it established black and gray lists for substances to be banned (black) or regulated by national authorities (gray). Cyanide and high-level radioactive waste, for example, were put on the black list. The London Convention applied only to waste dumped from ships, and thus did nothing to regulate waste discharged as liquids from pipelines.

Pathways of Pollution

Figure: *Septic river.*

There are many different ways to categorize, and examine the inputs of pollution into our marine ecosystems. Patin (n.d.) notes that generally there are three main types of inputs of pollution into the ocean: direct discharge of waste into the oceans, runoff into the waters due to rain, and pollutants that are released from the atmosphere.

One common path of entry by contaminants to the sea are rivers. The evaporation of water from oceans exceeds precipitation. The balance is restored by rain over the continents entering rivers and then being returned to the sea. The Hudson in New York State and the Raritan in New Jersey, which empty at the northern and southern ends of Staten Island, are a source of mercury contamination of zooplankton (copepods) in the open ocean. The highest concentration in the filter-feeding copepods is not at the mouths of these rivers but 70 miles south, nearer Atlantic City, because water flows close to the coast. It takes a few days before toxins are taken up by the plankton.

Pollution is often classed as point source or nonpoint source pollution. Point source pollution occurs when there is a single, identifiable, and localized source of the pollution. An example is directly discharging sewage and industrial waste into the ocean. Pollution such as this occurs particularly in developing nations. Nonpoint source pollution occurs when the pollution comes from ill-defined and diffuse sources. These can be difficult to regulate. Agricultural runoff and wind blown debris are prime examples.

Direct Discharge

Pollutants enter rivers and the sea directly from urban sewerage and industrial waste discharges, sometimes in the form of hazardous and toxic wastes.

Inland mining for copper, gold. etc., is another source of marine pollution. Most of the pollution is simply soil, which ends up in rivers flowing to the sea. However, some minerals discharged in the course of the mining can cause problems, such as copper, a common industrial pollutant, which can interfere with the life history and development of coral polyps. Mining has a poor environmental track record. For example, according to the United States Environmental Protection Agency, mining has contaminated portions of the headwaters of over 40% of watersheds in the western continental US. Much of this pollution finishes up in the sea.

Land Runoff

Surface runoff from farming, as well as urban runoff and runoff from the construction of roads, buildings, ports, channels, and harbours, can carry soil and particles laden with carbon, nitrogen, phosphorus, and minerals. This nutrient-rich water can cause fleshy algae and phytoplankton to thrive in coastal areas; known as algal blooms, which have the potential to create hypoxic conditions by using all available oxygen.

Polluted runoff from roads and highways can be a significant source of water pollution in coastal areas. About 75 percent of the toxic chemicals that flow into Puget Sound are carried by stormwater that runs off paved roads and driveways, rooftops, yards and other developed land.

Ship Pollution

Ships can pollute waterways and oceans in many ways. Oil spills can have devastating effects. While being toxic to marine life, polycyclic aromatic hydrocarbons (PAHs), found in crude oil, are very difficult to clean up, and last for years in the sediment and marine environment.

Figure: *A cargo ship pumps ballast water over the side.*

Oil spills are probably the most emotive of marine pollution events. However, while a tanker wreck may result in extensive newspaper headlines, much of the oil in the world's seas comes from other smaller sources, such as tankers discharging ballast water from oil tanks used on return ships, leaking pipelines or engine oil disposed of down sewers.

Discharge of cargo residues from bulk carriers can pollute ports, waterways and oceans. In many instances vessels intentionally discharge illegal wastes despite foreign and domestic regulation prohibiting such actions. It has been estimated that container ships lose over 10,000 containers at sea each year (usually during storms). Ships also create noise pollution that disturbs natural wildlife, and water from ballast tanks can spread harmful algae and other invasive species.

Ballast water taken up at sea and released in port is a major source of unwanted exotic marine life. The invasive freshwater zebra mussels, native to the Black, Caspian and Azov seas, were probably transported to the Great Lakes via ballast water from a transoceanic vessel. Meinesz believes that one of the worst cases of a single invasive species causing harm to an ecosystem can be attributed to a seemingly harmless jellyfish. *Mnemiopsis leidyi*, a species of comb jellyfish that spread so it now inhabits estuaries in many parts of the world. It was first introduced in 1982, and thought to have been transported to the Black Sea in a ship's ballast water. The population of the jellyfish shot up exponentially and, by 1988, it was wreaking havoc upon the local fishing industry. "The anchovy catch fell from 204,000 tons in 1984 to 200 tons in 1993; sprat from 24,600 tons in 1984 to 12,000 tons in 1993; horse mackerel from 4,000 tons in 1984 to zero in 1993." Now that the jellyfish have exhausted the zooplankton, including fish larvae, their numbers have fallen dramatically, yet they continue to maintain a stranglehold on the ecosystem.

Invasive species can take over once occupied areas, facilitate the spread of new diseases, introduce new genetic material, alter underwater seascapes and jeopardize the ability of native species to obtain food. Invasive species are responsible for about $138 billion annually in lost revenue and management costs in the US alone.

Ballast Water Discharge and the Environment

Ballast water discharges by ships can have a negative impact on the marine environment.

Cruise ships, large tankers, and bulk cargo carriers use a huge amount of ballast water, which is often taken on in the coastal waters in one region after ships discharge wastewater or unload cargo, and discharged at the next port of call, wherever more cargo is loaded. Ballast water discharge typically contains a variety of biological materials, including plants, animals, viruses, and bacteria. These materials often include non-native, nuisance, exotic species that can cause extensive ecological and economic damage to aquatic ecosystems, along with serious human health issues including death.

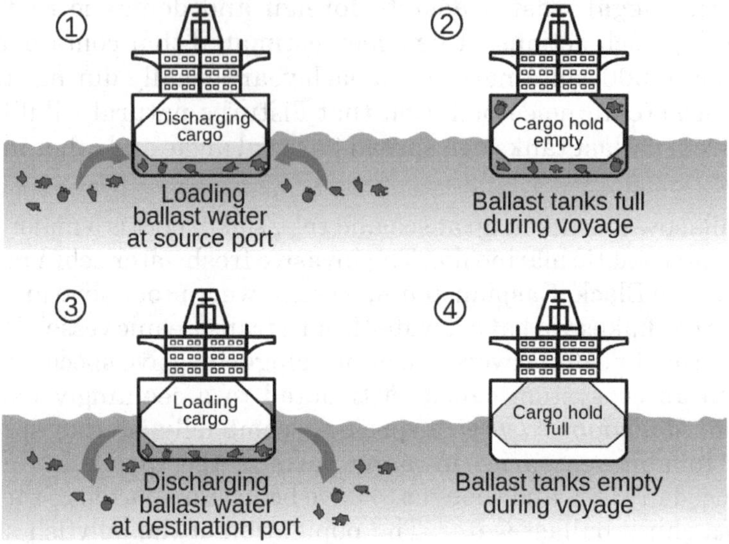

Figure: *Diagram showing the water pollution of the seas from untreated ballast water discharges*

Problematic Species

There are hundreds of organisms carried in ballast water that cause problematic ecological effects outside of their natural range. The International Maritime Organization list the ten most unwanted species as:

- Cholera *Vibrio cholerae* (various strains)
- Cladoceran Water Flea *Cercopagis pengoi*
- Mitten Crab *Eriocheir sinensis*
- Toxic algae (red/brown/green tides) (various species)
- Round Goby *Neogobius melanostomus*
- North American Comb Jelly *Mnemiopsis leidyi*
- North Pacific Seastar *Asterias amurensis*
- Zebra Mussel *Dreissena polymorpha*
- Asian Kelp *Undaria pinnatifida*
- European Green Crab *Carcinus maenas*

Ballast Water Issues by Country

New Zealand: The ballast tanks in New Zealand carry animals and plants that kill ecosystems. Ballast tanks are only used in cargo ships there. Ballast water is controlled under the Biosecurity Act 1993 in New Zealand.

Peru: A form of cholera, *Vibrio cholerae*, previously reported only in Bangladesh apparently arrived via ballast water in Peru in 1991, killing more than 10,000 people over the following three years.

United States: The zebra mussel, native to the Caspian and Black Seas arrived in Lake St. Clair in the ballast water of a transatlantic freighter in 1988 and within 10 years spread to all of the five neighbouring Great Lakes. The economic cost of this introduction has been estimated by the U.S. Fish and Wildlife Service at about $5 billion.

Ballast water discharges are believed to be the leading source of invasive species in U.S. marine waters, thus posing public health and environmental risks, as well as significant economic cost to industries such as water and power utilities, commercial and recreational fisheries, agriculture, and tourism. Studies suggest that the economic cost just from introduction of pest mollusks (zebra mussels, the Asian clam, and others) to U.S. aquatic ecosystems is more than $6 billion per year.

Congress passed the National Invasive Species Act in 1996 in order to regulate ballast water discharges. The law expired in 2002. The Environmental Protection Agency and Coast Guard both have some ongoing responsibilities for regulating pollution from ships, but ballast water discharges are not currently subject to permit requirements under the Clean Water Act.

Atmospheric Pollution

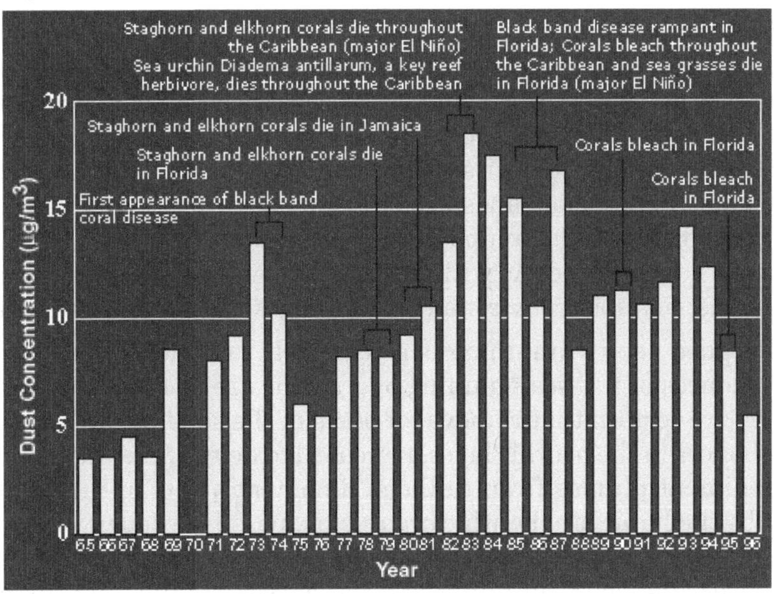

Figure: *Graph linking atmospheric dust to various coral deaths across the Caribbean Sea and Florida*

Another pathway of pollution occurs through the atmosphere. Wind blown dust and debris, including plastic bags, are blown seaward from landfills and other areas. Dust from the Sahara moving around the southern periphery of the subtropical ridge moves into the Caribbean and Florida during the warm season as the ridge builds and moves northward through the subtropical Atlantic. Dust can also be attributed to a global transport from the Gobi and Taklamakan deserts across Korea, Japan, and the Northern Pacific to the Hawaiian Islands. Since 1970, dust outbreaks have worsened due to periods of drought in Africa. There is a large variability in dust transport to the Caribbean and Florida from year to year; however, the flux is greater during positive phases of the North Atlantic Oscillation. The USGS links dust events to a decline in the health of coral reefs across the Caribbean and Florida, primarily since the 1970s.

Climate change is raising ocean temperatures and raising levels of carbon dioxide in the atmosphere. These rising levels of carbon dioxide are acidifying the oceans. This, in turn, is altering aquatic ecosystems and modifying fish distributions, with impacts on the sustainability of fisheries and the livelihoods of the communities that depend on them. Healthy ocean ecosystems are also important for the mitigation of climate change.

Deep Sea Mining

Deep sea mining is a relatively new mineral retrieval process that takes place on the ocean floor. Ocean mining sites are usually around large areas of polymetallic nodules or active and extinct hydrothermal vents at about 1,400 - 3,700 meters below the ocean's surface. The vents create sulfide deposits, which contain precious metals such as silver, gold, copper, manganese, cobalt, and zinc. The deposits are mined using either hydraulic pumps or bucket systems that take ore to the surface to be processed. As with all mining operations, deep sea mining raises questions about environmental damages to the surrounding areas

Because deep sea mining is a relatively new field, the complete consequences of full scale mining operations are unknown. However, experts are certain that removal of parts of the sea floor will result in disturbances to the benthic layer, increased toxicity of the water column and sediment plumes from tailings. Removing parts of the sea floor disturbs the habitat of benthic organisms, possibly, depending on the type of mining and location, causing permanent disturbances. Aside from direct impact of mining the area, leakage, spills and corrosion would alter the mining area's chemical makeup.

Among the impacts of deep sea mining, sediment plumes could have the greatest impact. Plumes are caused when the tailings from mining (usually fine particles) are dumped back into the ocean, creating a cloud of particles floating in the water. Two types of plumes occur: near bottom plumes and surface plumes. Near bottom plumes occur when the tailings are pumped back down to the mining site. The floating particles increase the turbidity, or cloudiness, of the water, clogging filter-feeding apparatuses used by benthic organisms. Surface plumes cause a more serious problem. Depending on the size of the particles and water currents the plumes could spread over vast areas. The plumes could impact zooplankton and light penetration, in turn affecting the food web of the area.

Types of Pollution

Acidification

Figure: *Island with fringing reef in the Maldives. Coral reefs are dying around the world.*

The oceans are normally a natural carbon sink, absorbing carbon dioxide from the atmosphere. Because the levels of atmospheric carbon dioxide are increasing, the oceans are becoming more acidic. The potential consequences of ocean acidification are not fully understood, but there are concerns that structures made of calcium carbonate may become vulnerable to dissolution, affecting corals and the ability of shellfish to form shells.

Oceans and coastal ecosystems play an important role in the global carbon cycle and have removed about 25% of the carbon dioxide emitted by human activities between 2000 and 2007 and about half the anthropogenic CO_2 released since the start of the industrial revolution. Rising ocean temperatures and ocean acidification means that the capacity of the ocean carbon sink will gradually get weaker, giving rise to global concerns expressed in the Monaco and Manado Declarations.

A report from NOAA scientists published in the journal Science in May 2008 found that large amounts of relatively acidified water are upwelling to within four miles of the Pacific continental shelf area of North America. This area is a critical zone where most local marine life lives or is born. While the paper dealt only with areas from Vancouver to northern California, other continental shelf areas may be experiencing similar effects.

A related issue is the methane clathrate reservoirs found under sediments on the ocean floors. These trap large amounts of the greenhouse gas methane, which ocean warming has the potential to release. In 2004 the global inventory of ocean methane clathrates was estimated to occupy between one and five million cubic kilometres. If all these clathrates were to be spread uniformly across the ocean floor, this would translate to a thickness between three and fourteen metres. This estimate corresponds to 500-2500 gigatonnes carbon (Gt C), and can be compared with the 5000 Gt C estimated for all other fossil fuel reserves.

Eutrophication

Eutrophication is an increase in chemical nutrients, typically compounds containing nitrogen or phosphorus, in an ecosystem. It can result in an increase in the ecosystem's primary productivity (excessive plant growth and decay), and further effects including lack of oxygen and severe reductions in water quality, fish, and other animal populations.

The biggest culprit are rivers that empty into the ocean, and with it the many chemicals used as fertilizers in agriculture as well as waste from livestock and humans. An excess of oxygen depleting chemicals in the water can lead to hypoxia and the creation of a dead zone.

Estuaries tend to be naturally eutrophic because land-derived nutrients are concentrated where runoff enters the marine environment in a confined channel. The World Resources Institute has identified

375 hypoxic coastal zones around the world, concentrated in coastal areas in Western Europe, the Eastern and Southern coasts of the US, and East Asia, particularly in Japan. In the ocean, there are frequent red tide algae blooms that kill fish and marine mammals and cause respiratory problems in humans and some domestic animals when the blooms reach close to shore.

Figure: *Polluted lagoon.*

In addition to land runoff, atmospheric anthropogenic fixed nitrogen can enter the open ocean. A study in 2008 found that this could account for around one third of the ocean's external (non-recycled) nitrogen supply and up to three per cent of the annual new marine biological production. It has been suggested that accumulating reactive nitrogen in the environment may have consequences as serious as putting carbon dioxide in the atmosphere.

One proposed solution to eutrophication in estuaries is to restore shellfish populations, such as oysters. Oyster reefs remove nitrogen from the water column and filter out suspended solids, subsequently reducing the likelihood or extent of harmful algal blooms or anoxic conditions. Filter feeding activity is considered beneficial to water quality by controlling phytoplankton density and sequestering nutrients, which can be removed from the system through shellfish harvest, buried in the sediments, or lost through denitrification. Foundational work toward the idea of improving marine water quality through shellfish cultivation to was conducted by Odd Lindahl et al., using mussels in Sweden.

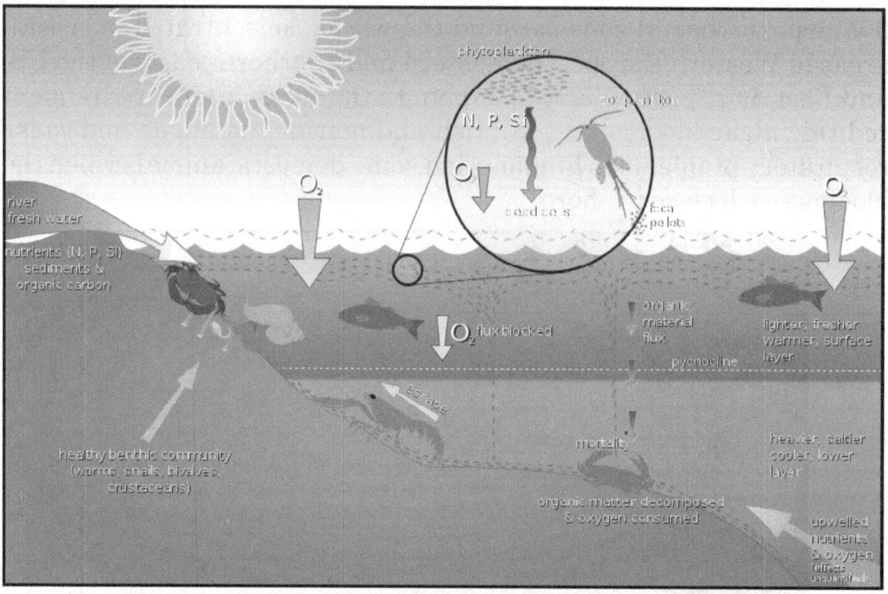

Figure: *Effect of eutrophication on marine benthic life*

Plastic Debris

Marine debris is mainly discarded human rubbish which floats on, or is suspended in the ocean. Eighty percent of marine debris is plastic - a component that has been rapidly accumulating since the end of World War II. The mass of plastic in the oceans may be as high as one hundred million metric tons.

Discarded plastic bags, six pack rings and other forms of plastic waste which finish up in the ocean present dangers to wildlife and fisheries. Aquatic life can be threatened through entanglement, suffocation, and ingestion. Fishing nets, usually made of plastic, can be left or lost in the ocean by fishermen. Known as ghost nets, these entangle fish, dolphins, sea turtles, sharks, dugongs, crocodiles, seabirds, crabs, and other creatures, restricting movement, causing starvation, laceration and infection, and, in those that need to return to the surface to breathe, suffocation.

Many animals that live on or in the sea consume flotsam by mistake, as it often looks similar to their natural prey. Plastic debris, when bulky or tangled, is difficult to pass, and may become permanently lodged in the digestive tracts of these animals. Especially when evolutionary adaptions make it impossible for the likes of turtles to reject plastic bags, which resemble jellyfish when immersed in water, as they have a system in their throat to stop slippery foods from otherwise escaping. Thereby blocking the passage of food and causing death through starvation or infection.

***Figure:** A mute swan builds a nest using plastic garbage.*

Plastics accumulate because they don't biodegrade in the way many other substances do. They will photodegrade on exposure to the sun, but they do so properly only under dry conditions, and water inhibits this process.

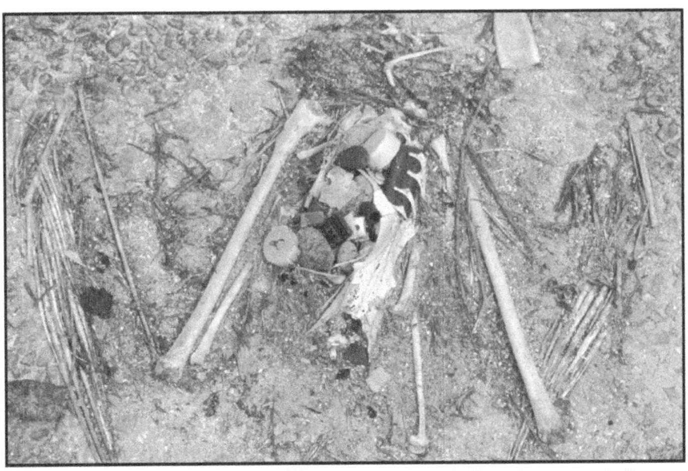

***Figure:** Remains of an albatross containing ingested flotsam*

In marine environments, photodegraded plastic disintegrates into ever smaller pieces while remaining polymers, even down to the molecular level. When floating plastic particles photodegrade down to zooplankton sizes, jellyfish attempt to consume them, and in this way the plastic enters the ocean food chain. Many of these long-lasting pieces end up in the stomachs of marine birds and animals, including sea turtles, and black-footed albatross.

Figure: *Marine debris on Kamilo Beach, Hawaii, washed up from the Great Pacific Garbage Patch*

Plastic debris tends to accumulate at the centre of ocean gyres. In particular, the Great Pacific Garbage Patch has a very high level of plastic particulate suspended in the upper water column. In samples taken in 1999, the mass of plastic exceeded that of zooplankton (the dominant animal life in the area) by a factor of six. Midway Atoll, in common with all the Hawaiian Islands, receives substantial amounts of debris from the garbage patch. Ninety percent plastic, this debris accumulates on the beaches of Midway where it becomes a hazard to the bird population of the island. Midway Atoll is home to two-thirds (1.5 million) of the global population of Laysan Albatross. Nearly all of these albatross have plastic in their digestive system and one-third of their chicks die.

Toxic additives used in the manufacture of plastic materials can leach out into their surroundings when exposed to water. Waterborne hydrophobic pollutants collect and magnify on the surface of plastic debris, thus making plastic far more deadly in the ocean than it would be on land. Hydrophobic contaminants are also known to bioaccumulate in fatty tissues, biomagnifying up the food chain and putting pressure on apex predators. Some plastic additives are known to disrupt the endocrine system when consumed, others can suppress the immune system or decrease reproductive rates. Floating debris can also absorb persistent organic pollutants from seawater, including PCBs, DDT and PAHs. Aside from toxic effects, when ingested some of these are mistaken by the animal brain for estradiol, causing hormone disruption in the affected wildlife.

Toxins

Apart from plastics, there are particular problems with other toxins that do not disintegrate rapidly in the marine environment. Examples of persistent toxins are PCBs, DDT, TBT, pesticides, furans, dioxins, phenols and radioactive waste. Heavy metals are metallic chemical elements that have a relatively high density and are toxic or poisonous at low concentrations. Examples are mercury, lead, nickel, arsenic and cadmium. Such toxins can accumulate in the tissues of many species of aquatic life in a process called bioaccumulation. They are also known to accumulate in benthic environments, such as estuaries and bay muds: a geological record of human activities of the last century.

Specific Examples

- Chinese and Russian industrial pollution such as phenols and heavy metals in the Amur River have devastated fish stocks and damaged its estuary soil.
- Wabamun Lake in Alberta, Canada, once the best whitefish lake in the area, now has unacceptable levels of heavy metals in its sediment and fish.
- Acute and chronic pollution events have been shown to impact southern California kelp forests, though the intensity of the impact seems to depend on both the nature of the contaminants and duration of exposure.
- Due to their high position in the food chain and the subsequent accumulation of heavy metals from their diet, mercury levels can be high in larger species such as bluefin and albacore. As a result, in March 2004 the United States FDA issued guidelines recommending that pregnant women, nursing mothers and children limit their intake of tuna and other types of predatory fish.
- Some shellfish and crabs can survive polluted environments, accumulating heavy metals or toxins in their tissues. For example, mitten crabs have a remarkable ability to survive in highly modified aquatic habitats, including polluted waters. The

farming and harvesting of such species needs careful management if they are to be used as a food.

- Surface runoff of pesticides can alter the gender of fish species genetically, transforming male into female fish.
- Heavy metals enter the environment through oil spills - such as the Prestige oil spill on the Galician coast - or from other natural or anthropogenic sources.
- In 2005, the 'Ndrangheta, an Italian mafia syndicate, was accused of sinking at least 30 ships loaded with toxic waste, much of it radioactive. This has led to widespread investigations into radioactive-waste disposal rackets.
- Since the end of World War II, various nations, including the Soviet Union, the United Kingdom, the United States, and Germany, have disposed of chemical weapons in the Baltic Sea, raising concerns of environmental contamination.

Underwater Noise

Marine life can be susceptible to noise or the sound pollution from sources such as passing ships, oil exploration seismic surveys, and naval low-frequency active sonar. Sound travels more rapidly and over larger distances in the sea than in the atmosphere. Marine animals, such as cetaceans, often have weak eyesight, and live in a world largely defined by acoustic information.

This applies also to many deeper sea fish, who live in a world of darkness. Between 1950 and 1975, ambient noise in the ocean increased by about ten decibels (that is a tenfold increase).

Noise also makes species communicate louder, which is called the Lombard vocal response. Whale songs are longer when submarine-detectors are on. If creatures don't "speak" loud enough, their voice can be masked by anthropogenic sounds.

These unheard voices might be warnings, finding of prey, or preparations of net-bubbling. When one species begins speaking louder, it will mask other species voices, causing the whole ecosystem to eventually speak louder.

According to the oceanographer Sylvia Earle, "Undersea noise pollution is like the death of a thousand cuts. Each sound in itself may not be a matter of critical concern, but taken all together, the noise from shipping, seismic surveys, and military activity is creating a totally different environment than existed even 50 years ago. That high level of noise is bound to have a hard, sweeping impact on life in the sea."

Marine Mammals and Sonar

Active sonar, the transmission equipment used on some ships to assist with navigation, may be detrimental to the health and livelihood of some marine animals. Research has recently shown that beaked and blue whales are sensitive to mid-frequency active sonar and move rapidly away from the source of the sonar, a response that disrupts their feeding and can cause mass strandings. Some marine animals, such as whales and dolphins, use echolocation or "biosonar" systems to locate predators and prey. It is conjectured that active sonar transmitters could confuse these animals and interfere with basic biological functions such as feeding and mating. Study has shown whales experience decompression sickness, a disease that forces nitrogen into gas bubbles in the tissues and is caused by rapid and prolonged surfacing. Although whales were originally thought to be immune to this disease, sonar has been implicated in causing behavioral changes that can lead to decompression sickness.

The impact of underwater sound can be reduced by limiting the sound exposure received by an animal. The maximum sound exposure level recommended by Southall et al. for cetaceans is 215 dB re 1 μPa^2 s for hearing damage. Maximum sound pressure level for behavioural effects is dependent on context (Southall et al.).

A great deal of the legal and media conflict on this issue has to do with questions of who determines what type of mitigation is sufficient. Coastal commissions, for example, were originally thought to only have legal responsibility for beachfront property, and state waters (three miles into sea). Because active sonar is instrumental to ship defense, mitigation measures that may seem sensible to a civilian agency without any military or scientific background can have disastrous effects on training and readiness. Navies therefore often define their own mitigation requirements.

Adaptation and Mitigation

Much anthropogenic pollution ends up in the ocean. The 2011 edition of the United Nations Environment Programme Year Book identifies as the main emerging environmental issues the loss to the oceans of massive amounts of phosphorus, "a valuable fertilizer needed to feed a growing global population", and the impact billions of pieces of plastic waste are having globally on the health of marine environments. Bjorn Jennssen (2003) notes in his article, "Anthropogenic pollution may reduce biodiversity and productivity of marine ecosystems, resulting in reduction and depletion of human marine food resources".

There are two ways the overall level of this pollution can be mitigated: either the human population is reduced, or a way is found to reduce the ecological footprint left behind by the average human. If the second way is not adopted, then the first way may be imposed as world ecosystems falter.

Figure: *Aerosol can polluting a beach.*

The second way is for humans, individually, to pollute less. That requires social and political will, together with a shift in awareness so more people respect the environment and are less disposed to abuse it. At an operational level, regulations, and international government participation is needed. It is often very difficult to regulate marine pollution because pollution spreads over international barriers, thus making regulations hard to create as well as enforce.

Without appropriate awareness of marine pollution, the necessary global will to effectively address the issues may prove inadequate. Balanced information on the sources and harmful effects of marine pollution need to become part of general public awareness, and ongoing research is required to fully establish, and keep current, the scope of the issues. As expressed in Daoji and Dag's research, one of the reasons why environmental concern is lacking among the Chinese is because the public awareness is low and therefore should be targeted. Likewise, regulation, based upon such in-depth research should be employed. In California, such regulations have already been put in place to protect Californian coastal waters from agricultural runoff. This includes the California Water Code, as well as several voluntary programs.

Similarly, in India, several tactics have been employed that help reduce marine pollution, however, they do not significantly target the problem.

In Chennai, sewage has been dumped further into open waters. Due to the mass of waste being deposited, open-ocean is best for diluting, and dispersing pollutants, thus making them less harmful to marine ecosystems.

Marine Debris

Marine debris, also known as marine litter, is human-created waste that has deliberately or accidentally been released in a lake, sea, ocean or waterway. Floating oceanic debris tends to accumulate at the centre of gyres and on coastlines, frequently washing aground, when it is known as *beach litter* or tidewrack. Deliberate disposal of wastes at sea is called *ocean dumping*. Naturally occurring debris, such as driftwood, are also present.

Figure: *Marine debris on the Hawaiian coast*

With the increasing use of plastic, human influence has become an issue as many types of plastics do not biodegrade. Waterborne plastic poses a serious threat to fish, seabirds, marine reptiles, and marine mammals, as well as to boats and coasts. Dumping, container spillages, litter washed into storm drains and waterways and wind-blown landfill waste all contribute to this problem.

Types of Debris

Researchers classify debris as either land- or ocean-based; in 1991, the United Nations Joint Group of Experts on the Scientific Aspects of Marine Pollution estimated that up to 80% of the pollution was land-based. A wide variety of anthropogenic artifacts can become marine debris; plastic bags, balloons, buoys, rope, medical waste, glass bottles and plastic bottles, cigarette lighters, beverage cans, polystyrene, lost fishing line and nets, and various wastes from cruise ships and oil rigs are among the items commonly found to have washed ashore. Six pack rings, in particular, are considered emblematic of the problem.

Eighty percent of marine debris is plastic. Plastics accumulate because they typically don't biodegrade as many other substances do. They photodegrade on exposure to sunlight, although they do so only under dry conditions, as water inhibits photolysis.

Ghost Nets

Fishing nets left or lost in the ocean by fishermen – ghost nets – can entangle fish, dolphins, sea turtles, sharks, dugongs, crocodiles, seabirds, crabs and other creatures. These nets restrict movement, causing starvation, laceration and infection, and, in animals that breathe air, suffocation.

Nurdles and Plastic Bags

Nurdles, also known as "mermaids' tears", are plastic pellets, typically under five millimetres in diameter, that are a major component of marine debris. They are a raw material in plastics manufacturing, and enter the natural environment when spilled. Weathering produces ever smaller pieces. Nurdles strongly resemble fish eggs.

Figure: *A handful of nurdles, spilt from a train in Pineville, Louisiana*

Plastic waste has reached all the world's oceans. This pollution harms an estimated 100,000 sea turtles and marine mammals and 1,000,000 sea creatures each year. Pelagic plastic pieces in the center of our ocean's gyres outnumber live marine plankton, and are passed up the food chain to reach all marine life.

Plastic shopping bags can clog digestive tracts when consumed and can cause starvation through restricting the movement of food, or by filling the stomach and tricking the animal into thinking it is full. A 1994 study of the seabed using trawl nets in the North-Western Mediterranean around the coasts of Spain, France and Italy reported mean concentrations of debris of 1,935 items per square kilometre. Plastic debris accounted for 77%, of which 93% was plastic bags.

Deep-sea Debris

Litter, made from diverse materials that are denser than surface water (such as glasses, metals and some plastics), have been found to spread over the floor of seas and open oceans, where it can become entangled in corals and interfere with other sea-floor life, or even become buried under sediment, making clean-up extremely difficult, especially due to the wide area of its dispersal compared to shipwrecks.

Research performed by MBARI found items including plastic bags below 2000m depth off the west coast of North America and around Hawaii.

Sources of Debris

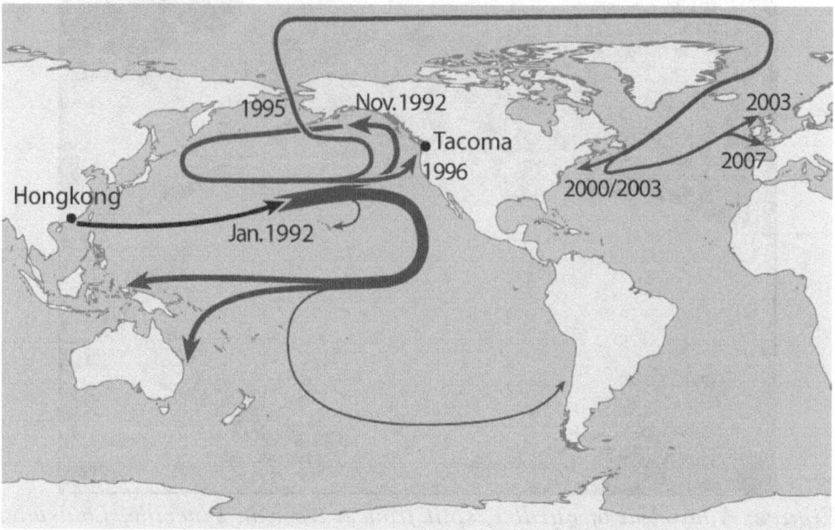

Figure: *Travel of the Friendly Floatees*

An estimated 10,000 containers at sea each year are lost by container ships, usually during storms. One famous spillage occurred in the Pacific Ocean in 1992, when thousands of rubber ducks and other toys went overboard during a storm.

The toys have since been found all over the world, providing a better understanding of ocean currents.

Similar incidents have happened before, such as when Hansa Carrier dropped 21 containers (with one notably containing buoyant Nike shoes). In 2007, MSC Napoli beached in the English Channel, dropping hundreds of containers, most of which washed up on the Jurassic Coast, a World Heritage Site.

In Halifax Harbour, Nova Scotia 52% of items were generated by recreational use of an urban park, 14% from sewage disposal and only 7% from shipping and fishing activities. Around four fifths of oceanic debris is from rubbish blown onto the water from landfills, and urban runoff.

In the 1987 Syringe Tide, medical waste washed ashore in New Jersey after having been blown from Fresh Kills Landfill. On the remote sub-Antarctic island of South Georgia, fishing-related debris, approximately 80% plastics, are responsible for the entanglement of large numbers of Antarctic fur seals.

Marine litter is even found on the floor of the Arctic ocean.

Great Pacific Garbage Patch

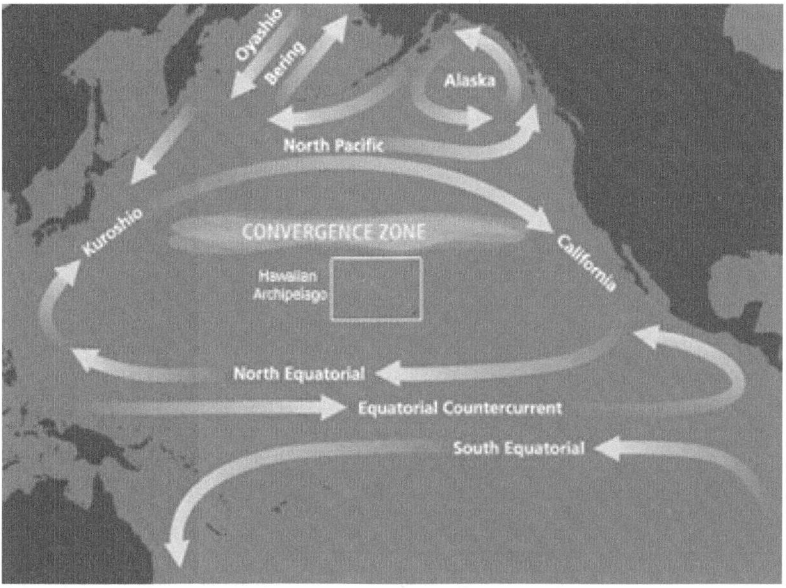

Figure: *The currents of the North Pacific Gyre spiral inwards, depositing debris in the convergence zone*

Once waterborne, debris becomes mobile. Flotsam can be blown by the wind, or follow the flow of ocean currents, often ending up in the middle of oceanic gyres where currents are weakest. The Great Pacific Garbage Patch is one such example of this, comprising a vast region of the North Pacific Ocean rich with anthropogenic wastes. Estimated to be double the size of Texas, the area contains more than 3 million tons of plastic. The gyre contains approximately six pounds of plastic for every pound of plankton. The oceans may contain as much as one hundred million tons of plastic.

Islands situated within gyres frequently have coastlines flooded by waste that washes ashore; prime examples are Midway and Hawaii. Clean-up teams around the world patrol beaches to attack this environmental threat.

Environmental Impact

Many animals that live on or in the sea consume flotsam by mistake, as it often looks similar to their natural prey. Bulky plastic debris may become permanently lodged in the digestive tracts of these animals, blocking the passage of food and causing death through starvation or infection. Tiny floating plastic particles also resemble zooplankton, which can lead filter feeders to consume them and cause them to enter the ocean food chain. In samples taken from the North Pacific Gyre in 1999 by the Algalita Marine Research Foundation, the mass of plastic exceeded that of zooplankton by a factor of six.

Figure: *A turtle trapped in a ghost net, an abandoned fishing net*

Toxic additives used in plastic manufacturing can leach into their surroundings when exposed to water. Waterborne hydrophobic pollutants collect and magnify on the surface of plastic debris, thus making plastic more deadly in the ocean than it would be on land. Hydrophobic contaminants bioaccumulate in fatty tissues, biomagnifying up the food chain and pressuring apex predators and humans. Some plastic additives disrupt the endocrine system when consumed; others can suppress the immune system or decrease reproductive rates.

Not all anthropogenic artifacts in the oceans are harmful. Iron and concrete typically do little damage to the environment as they generally sink to the bottom and become immobile, and can even provide scaffolding for artificial reefs. Ships and subway cars have been deliberately sunk for that purpose. Some organisms have adapted to live on floating plastic debris, allowing them to disperse all over the world and become invasive species in remote ecosystems.

Debris Removal

Techniques for collecting and removing marine (or riverine) debris include the use of debris skimmer boats *(pictured)*. Devices such as these can be used where floating debris presents a danger to navigation. For example, the US Army Corps of Engineers removes 90 tons of "drifting material" from San Francisco Bay every month. The Corps has been doing this work since 1942, when a seaplane carrying Admiral Chester W. Nimitz collided with a piece of floating debris and sank, costing the life of its pilot. Once debris becomes "beach litter", collection by hand and specialized beach-cleaning machines are used to gather the debris.

Figure: *Skimmer boat used to remove floating debris and trash from the Potomac and Anacostia rivers*

Elsewhere, "trash traps" are installed on small rivers to capture waterborne debris before it reaches the sea. For example, South Australia's Adelaide operates a number of such traps, known as "trash racks" or "gross pollutant traps" on the Torrens River, which flows (during the wet season) into Gulf St Vincent.

In lakes or near the coast, manual removal can also be used. Project AWARE for example promotes the idea of letting dive clubs clean up litter, for example as a diving exercise.

On the sea, the removal of artificial debris (i.e. plastics) is still in its infancy. However some projects have been started which used ships with nets (Kaisei and New Horizon) to catch some plastics, primarily for research purposes. Another method to gather artificial litter has been proposed by Boyan Slat. He suggested using platforms with arms to gather the debris, situated inside the current of gyres.

Plastic-to-fuel Conversion Strategy

The Clean Oceans Project (TCOP) promotes conversion of the plastic waste into valuable liquid fuels, including gasoline, diesel and kerosene, using plastic-to-fuel conversion technology developed by Blest Co. Ltd., a Japanese environmental engineering company. TCOP plans to educate local communities and create a financial incentive for them to recycle plastic, keep their shorelines clean, and minimize plastic waste.

Laws and Treaties

Ocean dumping is controlled by international law, including:

- The London Convention (1972) – a United Nations agreement to control ocean dumping
- MARPOL 73/78 – a convention designed to minimize pollution of the seas, including dumping, oil and exhaust pollution

European Law

In 1972 and 1974, conventions were held in Oslo and Paris respectively, and resulted in the passing of the OSPAR Convention, an international treaty controlling marine pollution in the north-east Atlantic Ocean. The Barcelona Convention protects the Mediterranean Sea. The Water Framework Directive of 2000 is a European Union directive committing EU member states to free inland and coastal waters from human influence. In the United Kingdom, the Marine and Coastal Access Act is designed to "ensure clean healthy, safe, productive and biologically diverse oceans and seas, by putting in place better systems for delivering sustainable development of marine and coastal environment".

United States Law

In 1972, the United States Congress passed the Ocean Dumping Act, giving the Environmental Protection Agency power to monitor and regulate the dumping of sewage sludge, industrial waste, radioactive waste and biohazardous materials into the nation's territorial waters. The Act was amended sixteen years later to include medical wastes. It is illegal to dispose of any plastic in US waters.

Ownership

Property law, admiralty law and the law of the sea may be of relevance when lost, mislaid, and abandoned property is found at sea.

Salvage law rewards salvors for risking life and property to rescue the property of another from peril. On land the distinction between deliberate and accidental loss led to the concept of a "treasure trove". In the United Kingdom, shipwrecked goods should be reported to a Receiver of Wreck, and if identifiable, they should be returned to their rightful owner.

Shoreline Morphodynamics

A shore or a shoreline is the fringe of land at the edge of a large body of water, such as an ocean, sea, or lake. In physical oceanography, a shore is the wider fringe that is geologically modified by the action of the body of water past and present, while the beach is at the edge of the shore, representing the intertidal zone where there is one. In contrast to a coast, a shore can border any body of water, while the coast must border an ocean; that is, a coast is a type of shore. The word shore is often substituted for coast where an oceanic shore is meant.

Figure: *The rocky shoreline of Taylor Head Provincial Park in Eastern Shore, Nova Scotia, Canada*

Shores are influenced by the topography of the surrounding landscape, as well as by water induced erosion, such as waves. The geological composition of rock and soil dictates the type of shore which is created.

Coast and Shore

Coasts mark the area where dry land meets oceans or other large bodies of water. They are one of the most active environments found on Earth. The coastal landscape is ever-changing. It reflects the conflicting processes of erosion (the gradual wearing away of Earth surfaces through the action of wind and water) and deposition (the accumulation and building up of natural materials). It is a landscape that is affected by the interactions of Earth, the Moon, and the Sun.

Coasts are among the most beautiful and inspiring landscapes on the planet, whether they are scenes of torrential storms or serene calm. Throughout recorded history, humans have sought to live on or near coasts. Although coasts account for only 10 percent of Earth's land surface, they serve as home to two-thirds of the world's human population.

The Shape of the Land

Coast and shore, coastline and shoreline, are commonly used in place of each other. Technically, they are different areas along a coastal landscape. The line that marks the boundary between water and land is the shoreline. It constantly fluctuates because of the regular action of waves and tides. The shore is that strip of ground bordering a body of water that is alternately covered or exposed by waves or tides. The boundaries of the shore are marked by the shoreline at its farthest seaward (low tide) and farthest landward (high tide).

Tide is the periodic rising and falling of water in oceans and other large bodies of water in response to the gravitational attraction of the Moon and the Sun upon Earth. Although the Sun is larger than the Moon, the Moon is closer to Earth and, therefore, its gravitational force is approximately 2.2 times greater. The gravitational pull of the Moon creates two types of tides: high and low. A tidal bulge occurs in the oceans on

the side of Earth nearest the Moon; at the same time, a second tidal bulge occurs on the opposite side of Earth. This second bulge forms because the force of the Moon's gravity pulls the solid body of Earth slightly away from the water on Earth's far side. These bulges are high tides. The areas between the tidal bulges experience low tide.

As Earth rotates, the tides move over its surface. It takes approximately 24 hours and 50 minutes for a given point on Earth to make a complete cycle relative to the Moon, and two lunar tidal cycles occur during this time. Thus, the average time between high tides is

12 hours and 25 minutes. This is a generalized explanation of tides. They do not move evenly and predictably over Earth's surface. Variations in the depth of the oceans and the distribution of landmasses combine with other factors to produce highly complex tidal behavior.

The coast and coastline begin where the shore ends at its high tide mark (farthest landward). The line between the coast and the shore at high tide is the coastline. The coast extends landward from the coastline to the first major change in terrain features, which may be miles inland. This could be a highland or a forest or some other type of terrain. Sometimes, the change between the coast and the adjacent terrain is not so distinct.

Coasts are generally classified into two types: emergent and submergent. Emergent coasts are those in which land formerly under water has gradually risen above sea level through geologic uplift of the land or has been exposed because of a drop in sea level. Currently, sea level around the world is rising by an average of 0.1 inch (0.25 centimeter) per year because glaciers and ice sheets are melting due to global warming (an increase in the world's temperatures thought to be caused, in part, by the burning of fossil fuels and the depletion of the ozone layer). So an emergent coast in the present-day is one that is rising on average more than 0.1 inch (0.25 centimeter) per year. Coasts along Scandinavia, New England, California, and Hawaii are examples of emergent coasts. Submergent coasts are those in which formerly dry land has been gradually flooded, either by land sinking or by sea level rising. Coasts along the southeast Atlantic and the Gulf of Mexico are examples of submergent coasts.

Coastal landscapes may be broadly divided into rocky cliffs and sandy beaches and dunes. All coasts experience a combination of erosion and deposition to varying degrees. Emergent coasts are typically dominated by cliffs or high, steep faces of rock. Because the land is rising in these areas, its landforms are subject to erosion. As waves break against a cliff, certain features are formed, depending on the hardness of the rock. Initially, wave action may cut an indentation, called a wave-cut notch, at the base of the cliff. When the notch becomes larger, rock in the cliff face above the notch loses support and falls into the water where it is broken up by the action of the waves. This process continues and the cliff slowly retreats inland. As it does so, a horizontal bench of rock remains beneath the waves at high tide where the cliff once stood. This feature is called a wave-cut platform. Over time, as the land continues to rise, this platform may be elevated and a new cliff face formed.

The Literary Landscape

"At the foot of this cliff a great ocean beach runs north and south unbroken, mile lengthening into mile. Solitary and elemental, unsullied and remote, visited and possessed by the outer sea, these sands might be the end or the beginning of a world. Age by age, the sea here gives battle to the land; age by age, the earth struggles for her own, calling to her defense her energies and her creations, bidding her plants steal down upon the beach, and holding the frontier sands in a net of grass and roots which the storms wash free."

Henry Beston, The Outermost House: A Year of Life on the Great Beach of Cape Cod, 1928.

In areas where cliff rock is alternately hard and soft, headlands and bays may form. A headland is an elevated area of hard rock that projects out into an ocean or other large body of water. When soft rock is eroded away between headlands, a curved inlet that holds a body of water known as a bay forms. Because of its location, a headland receives the brunt of

Figure: *Coastal features and landforms of both emergent and submergent coasts.*

Wave attack (wave energy is spread out and weaker in bays). Erosion by water and wind may create distinct features such as sea caves, sea arches, and sea stacks in a headland. Sea caves arise when waves hollow out weak areas of rock in headlands. Waves may then erode the cave through the headland, or caves on either side of the headland may meet. In either case, a sea arch is formed. Erosion of the arch continues until its top portion collapses, leaving a column of hard rock known as a sea stack standing detached from the sea cliff. Continual wave erosion eventually reduces the stack into a stump.

Beaches occur when sand, gravel, and other loose material are deposited by waves along a shore. A beach extends landward from the shoreline at low tide to the shoreline at high tide during storms, when waves are at their highest. In general, a beach is a sandy shore. Beaches are commonly divided into two zones: the foreshore zone and the back-shore zone. The foreshore zone is the area between the ordinary low tide mark and the ordinary high tide mark. The backshore zone is the area normally affected by waves only during a storm at high tide. Behind the back-shore zone may be cliffs, vegetation, or dunes created by winds moving sand from the beach. Commonly separating the two zones is a distinct mound of sand or gravel, called a berm, that runs parallel to the shoreline. It is created by the action of waves and tides.

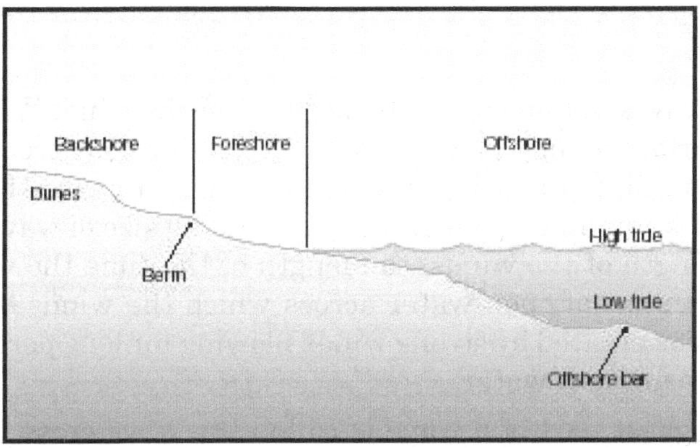

Figure: *Common elements of beach topography.*

Wave activity keeps sand and other loose material in constant motion. As a consequence, it can create other features along the shore, such as spits, bars, barrier islands, and tombolos. A spit is a long, narrow deposit of sand or gravel that projects from land into open water. Spits normally form at the mouth of a bay and curve inward. If a spit extends across the entire mouth of a bay, it is called a baymouth bar or bay barrier. A bar, commonly known as a sand bar, is a ridge or mound of sand or gravel that lies partially or completely underwater a short distance from and parallel to a beach. If more sand is deposited on the bar so that it rises above the normal high tide level, the bar becomes a barrier island. Barrier islands range in length from 1.8 to 62 miles (3 to 100 kilometers) and in width from 0.6 to 1.8 miles (1 to 3 kilometers). Separated from the shore by a shallow body of water known as a lagoon, a barrier island often helps protect the shore from the full force of waves. The East Coast of the United States from Massachusetts to

Florida is noted for its system of barrier islands. A mound of sand or other beach material that rises above the water to connect an offshore island to the shore or to another island is called a tombolo (pronounced TOM-beh-low).

Forces and Changes: Construction and Destruction

Coasts and shores are constantly changing. The shoreline moves with the waves and the tides. Rock is eroded away, and gravel and sand are deposited onshore, only to be swept back offshore. Storms batter coasts, and tides flood areas on a daily basis. The premiere forces that shape the coastal landscape, however, are waves. Breaking waves exert great force. A 10-foot (3-meter) wave can produce a force of 30 pounds per square inch (2.1 kilograms per square centimeter). In addition to the pressure exerted by their impact, waves erode by scouring rock cliffs and other coastal features with rock fragments they carry.

Waves

Most waves get energy and motion from the wind. Wind blowing over the surface of an ocean or other large water body creates friction along that surface, producing tiny ripples. Further pushed by the wind, these ripples combine and increase in size. The size of waves depends on the strength of the winds, the length of the time the winds blow, and the distance of open water across which the winds blow. Large waves may be created by strong winds blowing for long periods of time across large areas of water.

The highest part of a wave is called the wave crest. The lowest part of a wave between two crests is the wave trough. The vertical distance between the wave crest and the wave trough is the wave height. The horizontal distance between two wave crests or troughs is the wavelength. As a wave travels across an ocean or other body of water, the water particles in the wave move in circular patterns, in loops. These loops extend down underneath the surface of the water only one-half the distance of the wavelength. Water beneath that is not disturbed by wave motion. As the wave form advances across the surface, its energy moves forward, not the water itself. The water particles in the loops essentially return to their original position after the wave has passed.

As a wave enters shallow water near a shoreline, the lower loops in the wave begin to drag on the bottom. This causes the wave to slow down. As it does so, its wavelength decreases and its height increases. When the wave can no longer support its height, the wave breaks. Most of the energy it has carried across the ocean is then transferred into the churning, turbulent water known as surf. The powerful surf either crashes into a cliff or runs up a beach. The water moving up the beach is known as swash; the return flow toward the ocean is called the backwash. The speed of the swash is greater than that of the backwash.

Angled Waves and Drifting Sand

Most of the sand and other sediments making up a beach come from weathered and eroded rock from the mainland that is deposited by rivers at the coast. Material eroded from cliffs at the head of a beach itself may

Shoreline Processes and the Evolution of Coastal Landforms

Water Waves Are Another Agent Of An Erosion, Transportation And Deposition Of Sediments. Along the shores of oceans and lakes waves break against the land building it up in some places (depositing sediments) and tearing it down in others (eroding).

The Energy of the Waves Comes from the Wind

Energy is used to:	ERODETRANSPORT SEDIMENTS DEPOSIT

WAVE Alternating rise and fall of the water surface, produced by the flow of wind across the water surface. Small local differences in air pressure create ondulations in the water surface. It's important to understand how waves travel and move sediments.

Waves are described or characterized by

WAVELENGHT	Distance between identical parts of the wave.
CREST	Highest portion of the wave with respect to the horizontal
THROUGH	Lowest portion of the wave with respect to the horizontal
PERIOD	Time it takes identical parts of successive waves to pass a fixed reference point
HEIGHT	Vertical distance between the crest and the through. It is a key factor in determining wave energy OPEN OCEAN 2–5 m HURRICANE 15 m or more HIGHEST EVER 34 m (112 ft) Waves travel between 30-90 km/hour

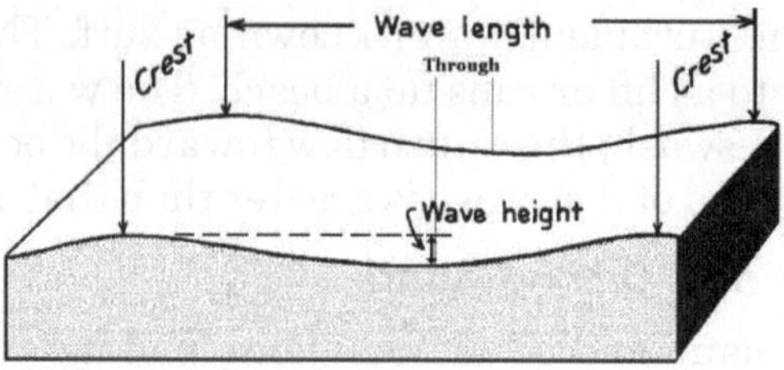

Tsunami: Results from a sudden displacement of seafloor (earthquake; fault; the collapse of a caldera) Wave height in open sea; 1 m or so. Wavelength in open sea; 100 to 200 km.

Movement in Water in a Wave

A particle of water moves in an orbit or circular path as the waves passes. The particle essentially returns to its original position after the wave has passed.

AT THE SURFACE	The diameter of the orbital path is equal to the height of the wave.
BELOW THE SURFACE	The orbit decreases in size until the motion is essentially gone at a depth equal to ½ of the wavelength. This is the WAVE BASE

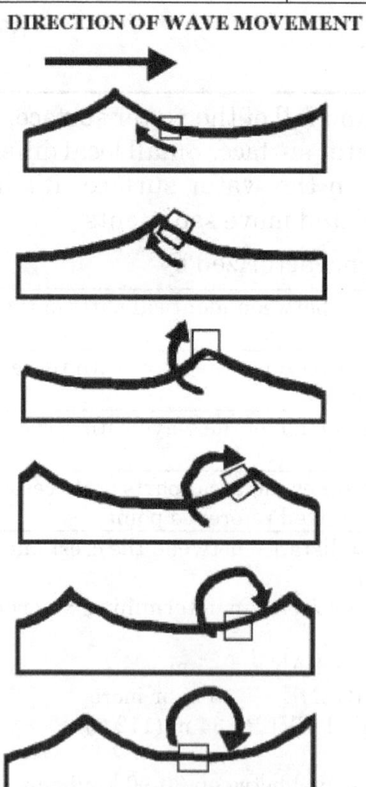

The motion of a water particle as a wave advances is indicated by the movement of a floating object. As the wave advances (left to right) the object is lifted to the crest and returns to the through. The wave form advances, but water particles move in an orbit returning to their original position

Wave Refraction

Most waves do not strike the shore directly, but rather meet it at an angle. When a wave approaches the coast in this manner, one end of the wave encounters shallow water first and slows down, while the rest of the waves is still in deeper water. Thus the part of the wave close to the shore slows down while the remainder continues to advance at a con tact speed. As a result the wave close to the shore bend. This process is called REFRACTION

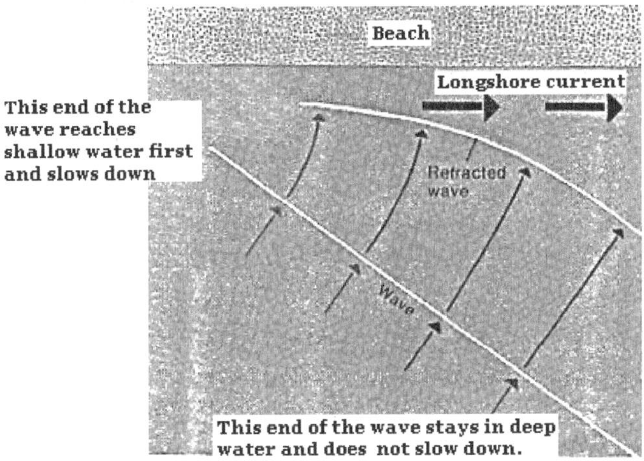

Wave refraction changes the wave direction, bending the wave so it becomes more parallel to shore. The angled approach of waves to shore sets up a longshore current to the shoreline

Wave Erosion

Occurs at sea level. Undercuts exposed bedrock. Waves break directly o the shoreline, and thus expend the greatest part of their energy in eroding land.

Features Caused by Deposition of Sediments

WAVE-CUT CLIFF	***Formed by waves pounding against a rocky coastline***
WAVE-CUT TERRACE or PLATFORM	Formed at the foot of a cliff. Formed by wave erosion pushing the Wave-cut cliff steadily back.
Features resulting from differential erosion of weaker sections of the rock	
SEA NOTCHS SEA CAVES	Wave action may hollow out cavities
SEA ARCHS	When erosion, cuts through a headlands
SEA STACKS	The collapse of the roof of a sea arch leaves an isolated mass of rock, in the front of the cliff.

Since the surging water of the breaking waves must cross this terrace before reaching the cliff, it looses a certain amount of energy through turbulence and friction. So the farther the cliff retreats the less effective are the waves eroding the cliff. If the sea level remains constant, the retreat of the cliff becomes slower and slower.

Erosional Features

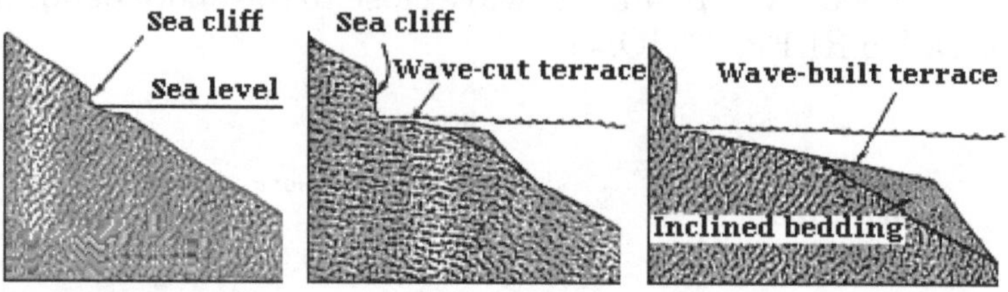

Cross section showing erosional and depositional features that result from wave action (vertical highly exaggerated

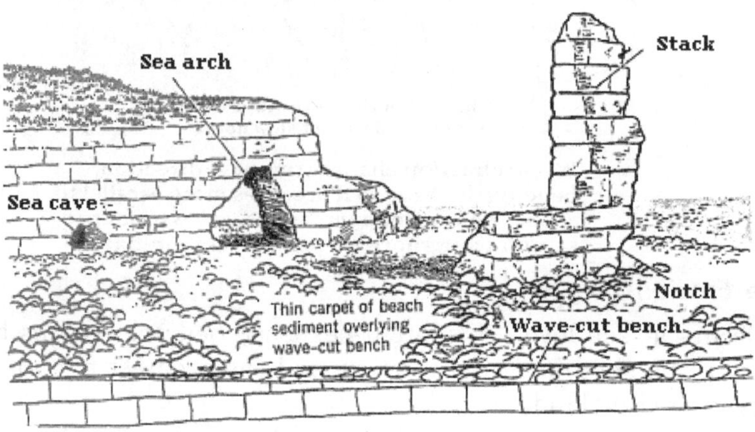

Minor erosional features along a rocky coast, seen at low tide. Surf hollows out sea cave in more erodible part of bedrock. Cave through headland becomes a sea arch. Surf tears away parts of bedrock, leaving isolated stack as an "island" on wave-cut platform (or bench)

Sediment Transport

Sediment is transported along the shore in a ZIGZAG pattern during the wash-backwash flow of the water. After the waves has washed up as far as it can go, the water returns to se by running down the beach face by the shortest possible route; straight downhill to the shoreline not back along the oblique route it came up (refraction). The net effect of this motion is to move the sand along the beach face. This transportation process is called LONG-SHORE DRIFT or CURRENT.

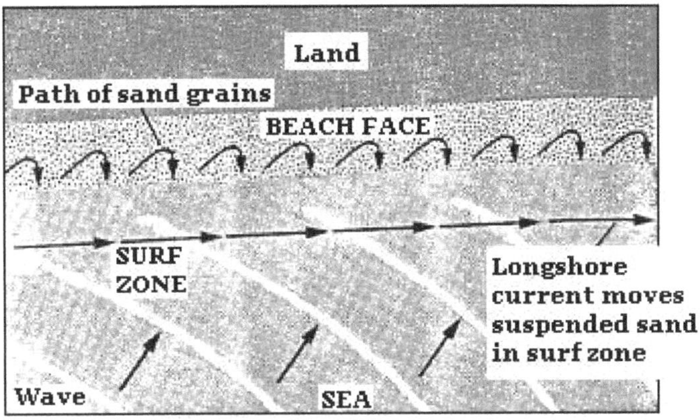

Features Caused by Deposition of Sediments

BEACH	Any strip of sediment that extends from the low-water line inland to a cliff or zone of permanent vegetation. Built of material eroded by waves from the headlands, and material brought down by rivers that carry the products of weathering and erosion from the land masses. Areas that are protected from the full force of water waves. It's continually modified by wave and current erosion.
SPIT	A narrow ridge or embankment of sediment forming a finger-like projection from the shore into the open ocean. Typically develops when the sediment being carried by long-shore drift is deposited where water becomes deeper, such as the mouth of a bay.
BAYMOUTH BAR or BAY BARRIER	A ridge, commonly exposed, of sand extending partially
TOMBOLO TIED ISLAND	Beach of sand or sand that connects two islands or an island with mainland.
BARRIER ISLAND (Atlantic City, NJ; Miami Beach, FL; and Galveston, TX)	Breached baymouth bar. Most common North America feature of the East coast. 1300 km from Long Island to Florida and the Gulf Coast (295) First line of defense against storm waves and hurricane.
LAGOONS	Isolated bays

Depositional Features

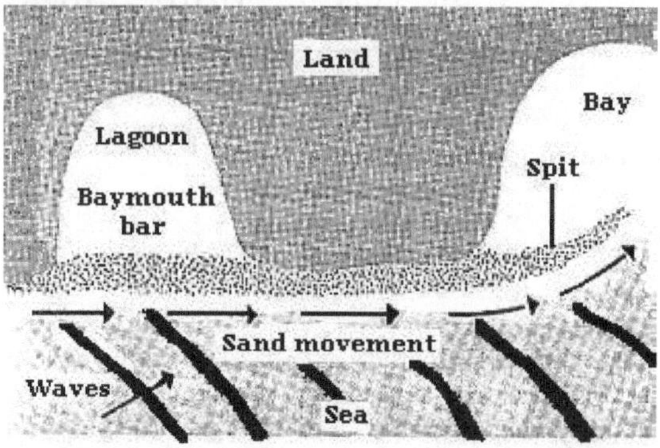

Longshore drift of sand can form spits and baymouth bars.

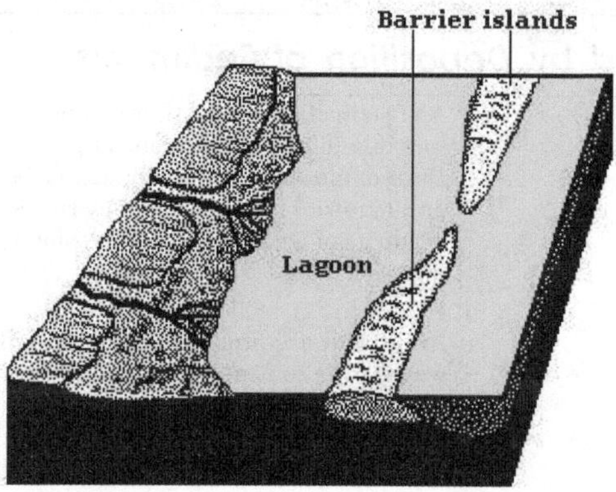

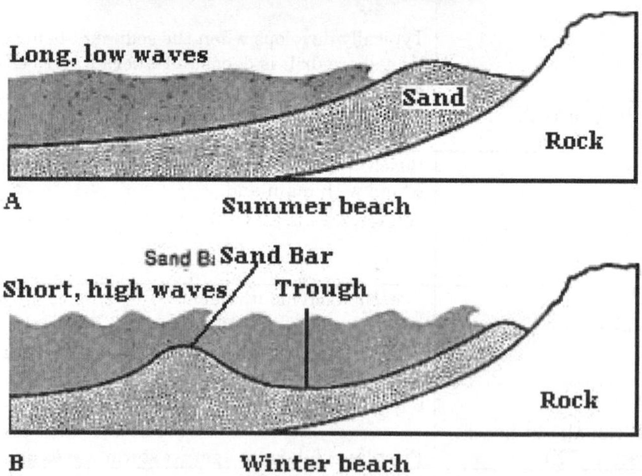

Seasonal cycle of a beach caused y differing waves types. (A) Summer beach. (B) Winter beach. Waves may break once on winter sandbar, then re-form and break again beach face.

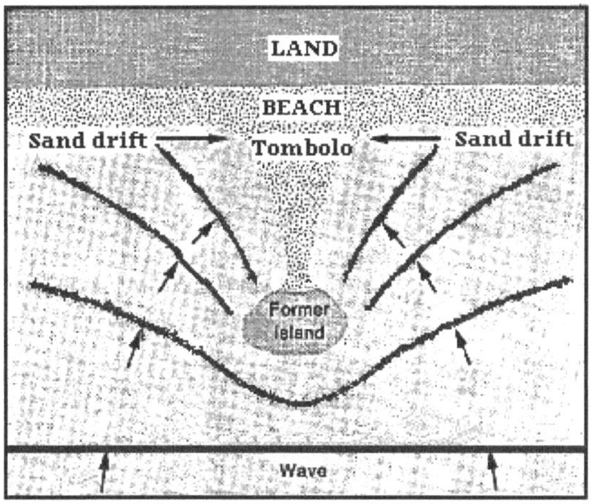

A tombolo connecting a former island to the mainland, forms by sand drift resulting from wave refraction caused by the island. Some sand may also be supplied by erosion of the island.

Main Types of Coastlines

SUBMERGENT	Flooded	Sea-level is relatively raised
EMERGENT	Exposed	Sea level is relatively lowered.

Two Main Types of Coastline

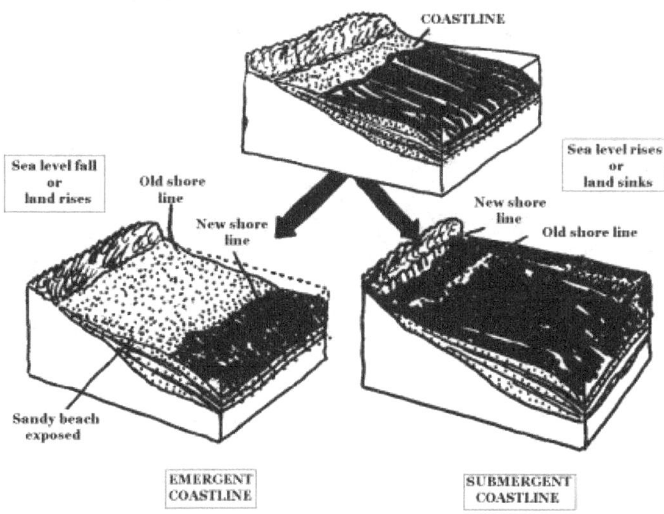

Submergent Coastlines

Previously dry land that has been recently drowned either by land subsidence or a rise in sea level. Their appearance is controlled by the erosional processes prior to the rise of the sea level. Irregular coastlines.

EROSIONAL PROCESS PRIOR TO THE RISE OF SEA	***TYPE /Features***
STREAM EROSION	ESTUARIES They extend inland Flooded stream valley V-shaped passages. Fresh-water from the river mixes with sea-water (brackish water) Protected environment. Reach marine life
GLACIAL EROSION	FJORDS They extend inland. Flooded glacial valley Forms very steep passages (U-shaped). Deep long narrow arm of he sea surrounded by high rocky cliffs or mountainous slopes Very rugged shoreline.

Submergent Coast Map (Estuary)

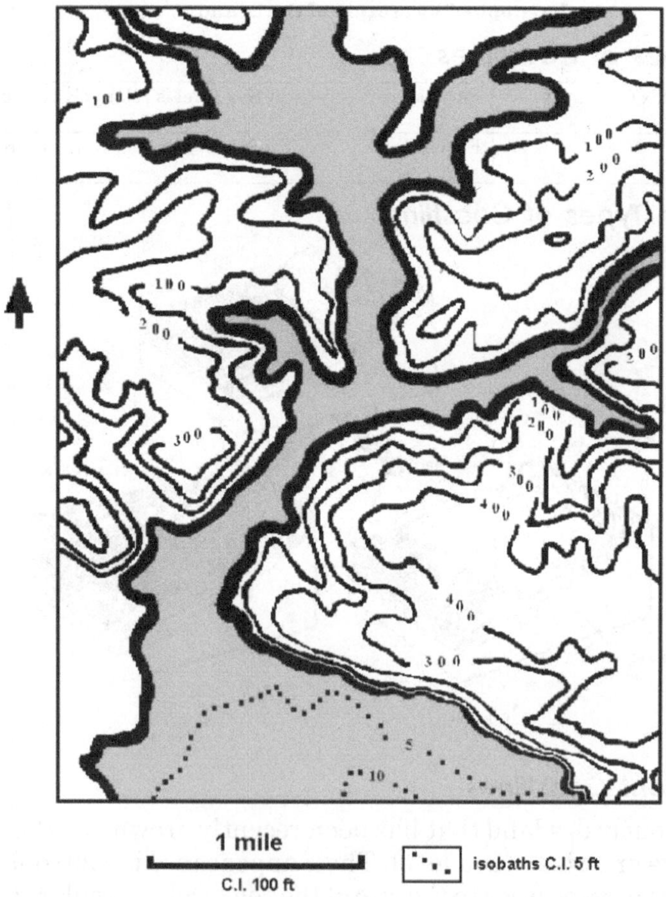

Submergent Coast Map (Fjord) Map

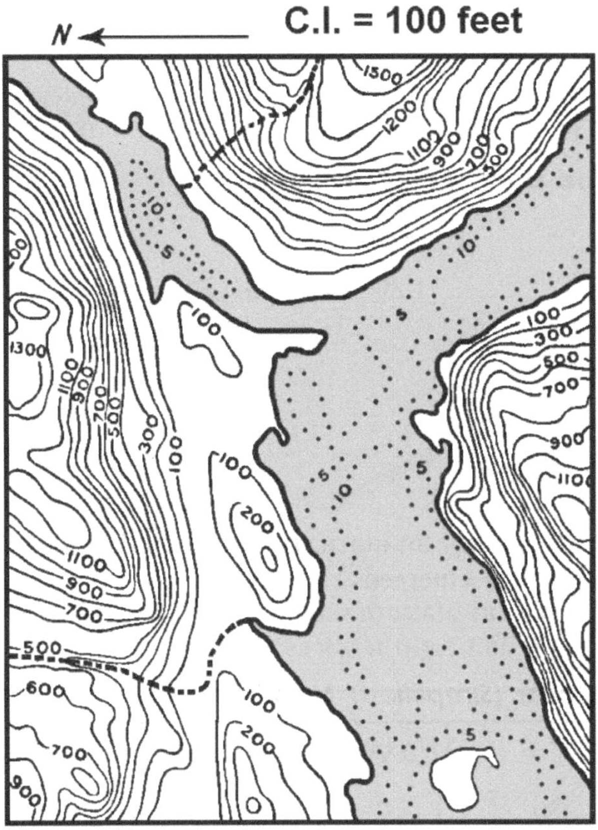

Delta: A delta is a thick accumulation of sediments at the mouth of a river.

As the shoreline is built has built outward, the river has shifted from one position to another.

This deposition has made the area so flat that the river has commonly split into a number of channels, each of which carries a part of the river water into the ocean (The Gulf). These are called DISTRIBUTARIES (PASSES).

Part of the sediment brought down b the distributaries is accumulated in the form of offshore islands (sand bars).

Waves and tides will modify and remove some of the delta front.

Mississippi River Delta: Each year the river discharges water into the Gulf at an average rate of 15,360 m3 per second with 684,528,300 tons of sediment (at the active Balize Delta Lobe). Dariry average discharge of 1-1.5 million tons of sediments)

Emergent Coastline: Land formed under water that recently has become dry land, either by uplift of the land or drop of the sea level.

Generally Straight

Characterized by the presence of extensive elevated Wave-Cut Platforms, Terraces or Platforms and Wave cut cliffs.

Erosion is the Dominant Agent

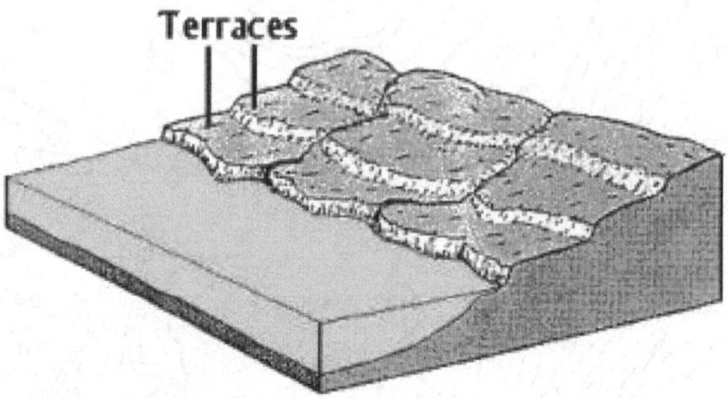

Block diagram illustrating characteristics of recently emergent coasts. Previous wave-cut platforms are elevated above sea level and form terraces

Emergent Coast (Simplified) Map

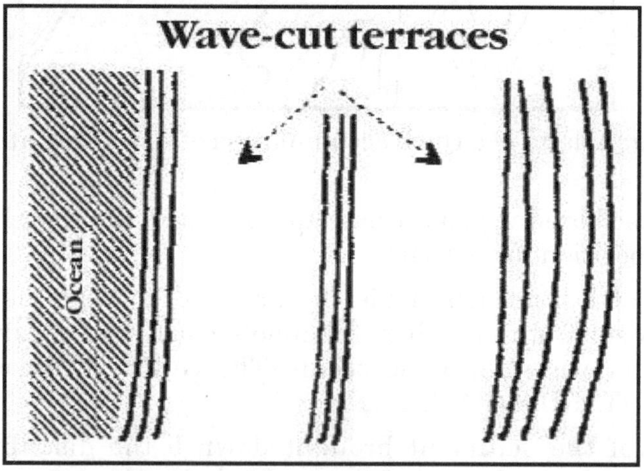

Stages Of Coast Modification (Submergent Coastlines)

When waves break against headlands most of their energy dissipates. Therefore the waves inside adjacent bays have lower energy and as a result sediments are deposited in the bays. As headlands erode and bays fill with sediment, an irregular coastline eventually becomes smooth

EARLY	Erosional features are dominant Landforms: sea cliffs and stacks, wave-cut platforms, etc. Depositional features not common. Highly irregular coastline
MIDDLE	Depositional features begin to appear; beaches, spits. Broad wave-cut platforms are common
LATE	Depositional features are dominant. Landforms: extended spits, Baymouth bars, beaches, Lagoons, etc. Reduced headlands. Nearly linear shorelines.

The Effects of Coastal Erosion on Shoreline Features

Shorelines constantly change due to the pounding of waves that erode coastal structures. Learn about shoreline features that are formed through coastal erosion, such as wave-cut cliffs, wave-cut platforms, sea arches and sea stacks.

The Shoreline

If you visited the beach as a kid and then went back with your children 30 years later, chances are the beach and shoreline would look different. This is because coastal areas are constantly changing in form and shape due to the impact of waves.

The pounding of waves is constantly eroding the shoreline, and this creates some interesting shoreline features. In this lesson, you will learn about the effects of coastal erosion, and how the relentless pounding of waves carves out unique features, such as cliffs, platforms, sea arches and sea stacks.

Coastal Erosion and Headlands

Coastal erosion, which is the wearing away of coastal land or beaches, is mainly caused by the impact of waves along the shoreline. This is accentuated during storms when waves are large and crash on shore with a lot of energy.

If a coastal structure, such as a headland, which is a narrow strip of land that projects out into a body of water, contains cracks or crevices, the waves can force water into the openings, dislodging land and rock fragments. Rocky structures along the coast are also subjected to erosion caused by sand and rocks stirred up by and carried within the waves. As the sand and rocks crash into the coast, they grind away land and rock.

Wave Refraction

Waves do not typically strike the shore head on. Wave refraction is the bending of waves as they travel toward the shallow waters of the shore. When waves move from the deep ocean waters to the shallow waters of the shore, the part of the wave closest to the shore hits bottom and slows first.

However, the part of the wave that is still in deep water proceeds at full speed. This contrast in speed causes a bending of the wave crest. As the wave approaches the coast, it is almost parallel to the shore. Because of wave refraction, the energy carried by waves pummels the sides and ends of shoreline protrusions, such as headlands leading to greater erosion.

Wave-Cut Cliffs and Wave-Cut Platforms

Shorelines are carved over time, leaving unique and oftentimes spectacular shoreline features. A wave-cut cliff is a steep rock face created by the eroding action of waves. As waves continue to hit the supporting rocks of the cliffs, a notch is carved into the cliff face. Rocks overhanging the erosion weaken and crumble, causing the cliff to retreat farther inland.

This leaves behind a somewhat flat area found at the base of a wave-cut cliff, called a wave-cut platform. These may be seen during times of low tide when the overlying water recedes. These platforms are sometimes referred to as coastal benches because when they are exposed during periods of low tide, they resemble large benches made of flat rock.

Sea Arches and Sea Stacks

As mentioned earlier, headlands that project into the ocean are savagely attacked by refracted waves. The waves erode the rocky headlands and chip away at any fractured rocks. This can lead to the creation of a sea arch, which is a natural hole carved through a headland, due to erosion by waves. Sea arches can be found along many coasts and each is uniquely beautiful.

Ecosystem of a Shoreline

Shoreline ecosystems happen where water meets land. Considering that water covers 75 percent of the planet, this area might appear extensive, but in reality it comprises a narrow space. Despite this fact, much life occurs around shorelines, and the ecosystems that develop there teem with biodiversity. Shorelines can be freshwater, saltwater or — where rivers meet the ocean — a mix of the two, which is called brackish water.

Freshwater Shorelines

A freshwater shoreline, the area directly surrounding a lake, encompasses the shallow area near shore as well as the area on land adjoining the water. Plants form the basis of the ecosystem, and in the water emergent plants dominate. Examples include water lilies, sedges, and arrow arum. These plants provide shelter and food to many different insects and small fish and also fertile hunting grounds for larger predators such as bass, pike, snapping turtles and wading birds. On the shore, willows and other water-loving trees grow and provide shelter and nesting places for birds. Raccoons and other opportunist omnivores feed in the shallow water, consuming crustaceans, fish, mollusks, frogs and toads, and other shoreline animals and plants.

Estuary Ecosystems

The ocean exerts a strong influence in brackish river estuary ecosystems. Estuaries run by the rhythm of the tides: when the tide comes in, the water will run upstream, and when it goes out, the water will run downstream. Salt marshes, the main type of shoreline ecosystem in estuaries, serve as nurseries of the ocean and have some of the highest levels of biodiversity in the world. Salt-tolerant grasses such as cord grass form the basis of the ecosystem. They die in the winter and provide food for a multitude of saltwater and freshwater animals.

Dune Ecosystem

Sand dunes, one of the most common shoreline types, skirt the edges of oceans and large lakes in many locations around the world. Dunes form when wind blows sand inland, where plants such as beach grass or sea grape trap the sand and it begins to pile up, creating a hill or sand dune. Although dunes may look relatively empty, many species of plants and animals inhabit them. Insects thrive in the dry grasses where birds and spade foot toads prey upon them. Shorebirds such as plovers and killdeer nest in low dunes. Because of high winds and tides, dunes are not permanent structures but constantly shift, move and change shape.

Mangrove Ecosystem

Mangrove swamps, another of the more common ocean shoreline ecosystems, exist across the world in tropical or sub-tropical climates. Mangroves build up shoreline and protect inland areas from storm damage. The roots of mangrove trees trap mud, sand, dirt and floating debris, and fish and other ocean animals shelter in the tangled roots. This attracts predators such as small sharks, crocodiles, pelicans and wading birds. As soil builds up, different mangroves take over, and eventually the area converts to land and the shoreline moves further into the ocean. Mangrove trees reproduce and spread by dropping large seeds into the water, where the current carries them to other locations.

Beach Morphodynamics

Beach morphodynamics refers to the study of the interaction and adjustment of the seafloor topography and fluid hydrodynamic processes, seafloor morphologies and sequences of change dynamics involving the motion of sediment. Hydrodynamic processes include those of waves, tides and wind-induced currents.

While hydrodynamic processes respond instantaneously to morphological change, morphological change requires the redistribution of sediment. As sediment takes a finite time to move, there is a lag in the morphological response to hydrodynamic forcing. Sediment can therefore be considered to be a time-dependent coupling mechanism. Since the boundary conditions of hydrodynamic forcing change regularly, this may mean that the beach never attains equilibrium.

This systems approach to the coast was developed by Wright and Thom in 1977.

Examples of beach morphodynamic processes include the formation of beach cusps (on a very small scale), intertidal bars and log-spiral (Yasso 1965)/crenulate (Silvester 1980) embayments. Morphodynamic processes exhibit positive and negative feedbacks (such that beaches can, over different timescales, be considered to be both self-forcing and self-organised systems), nonlinearities and threshold behaviour.

According to their dynamic and morphological characteristics, exposed sandy beaches can be classified into several morphodynamic types (Short & Wright 1983, Short 1996). There is a large scale of morphodynamic states, this scale ranges from the dissipative state to the reflective extremes. Dissipative beaches are flat, have fine sand, incorporating waves that tend to break far from the intertidal zone and dissipate force progressively along wide surf zones. Reflective beaches are steep, and are known for their coarse sand; they have no surf zone, and the waves break brusquely on the intertidal zone.

Depending on beach state, near bottom currents show variations in the relative dominance of motions due to: incident waves, subharmonic oscillations, infragravity oscillations, and mean longshore and rip currents. On reflective beaches, incident waves and subharmonic edge waves are dominant. In highly dissipative surf zones, shoreward decay of incident waves is accompanied by shoreward growth of infragravity energy; in the inner surf zone, currents associated with infragravity standing waves dominate. On intermediate states with pronounced bar-trough (straight or crescentic) topographies, incident wave orbital velocities are generally dominant but significant roles are also played by subharmonic and infragravity standing waves, longshore currents, and rips. The strongest rips and associated feeder currents occur in association with intermediate transverse bar and rip topographies.(Virginia Institute of Marine Science- Scopus 460).

With respect to beach morphodynamics, a spectrum of morphodynamic states has been developed, ranging from reflective beaches at one end to dissipative beaches at the other. Reflective beaches are typically steep in profile with a narrow shoaling and surf zone, composed of coarse sediment, and characterised by surging breakers. Coarser sediment allows percolation during the swash part of the wave cycle, thus reducing the strength of backwash and allowing material to be deposited in the swash zone. At the other end of the continuum of beach states, dissipative beaches are wide and flat in profile, with a wide shoaling and surf zone, composed of finer sediment, and characterised by spilling breakers. Transitions between beach states are often caused by changes in wave energy, with storms causing reflective beach profiles to flatten (offshore movement of sediment under steeper waves), thus adopting a more dissipative profile.

Morphodynamic processes are also associated with other coastal landforms, for example spur and groove topography on coral reefs and tidal flats in infilling estuaries.

Chapter 3

Longterm Coastal Development

Coastal areas are commonly defined as the interface or transition areas between land and sea, including large inland lakes. Coastal areas are diverse in function and form, dynamic and do not lend themselves well to definition by strict spatial boundaries. Unlike watersheds, there are no exact natural boundaries that unambiguously delineate coastal areas.

Geologically, continental margins are of two types: active margins where the edge of a continent happens to be at the edge of an oceanic plate (e.g. the west coast of South America); and inactive margins where the transition from continental lithosphere to oceanic lithosphere is within a plate rather than at a plate edge (e.g. the Atlantic). Coastal areas are therefore characterized by the vertical accretion of near-shore land. This depends on several factors: sediment supply from rivers or from the sea; the width of the shelf, or the proximity of a submarine canyon through which currents remove sediments; and the strength of longshore currents and incidence of cyclones, both of which transport and redistribute sediments along the coast. Sedimentation is the major geological activity that shapes coasts, but human-induced land subsidence is having an increasing impact on coastal morphology.

Nevertheless, for management purposes, a variety of landwards and seawards boundaries, ranging from fairly narrow and precise ones to much broader and more nebulous ones, have been utilized around the world. Management boundaries are pragmatic, being influenced by the geographic scope of relevant management concerns, including biophysical, economic, social, institutional and organizational aspects. Therefore, the boundaries of a coastal area may change over time for management purposes, as the issues to be faced become more extensive or complex and require more far-ranging solutions.

It has been suggested that a distinction be made between the terms 'coastal zone' and 'coastal area'. The term 'coastal zone' would refer to the geographic area defined by the enabling legislation for coastal management, while 'coastal area' would be used more broadly to refer to the geographic area along the coast that has not yet been defined as a zone for management purposes. In the United States, earlier attention to management in the coastal zone gave rise to the term 'coastal zone management' (CZM). As experience was gained of the multifaceted character of many of the issues and the consequent need to adopt an holistic approach to management, the term was revised to 'integrated coastal zone management' (ICZM). This term continues to be used by many authorities, including the World Bank.

The multifaceted approach to the management of coastal resources has become known as integrated coastal management (ICM). Pernetta and Elder have described it as meaning 'the process of combining all aspects of the human, physical and biological aspects of the coastal zone within a single management framework'. However, they have preferred the term 'holistic coastal management' to emphasize that 'careful planning and management of all sectoral activities simultaneously will result in greater overall benefits than pursuing sectoral development plans independently of one another.'

In practice, laws concerning coastal management seldom unambiguously or precisely define the coastal zone. Thus, the boundaries of the relevant management area can, and usually do, change over time without regard to the enabling legislation. In addition, few nations have comprehensive coastal zone management policies. As a result, different coastal areas within the same nation can fall under the jurisdiction of different coastal management plans and their boundaries can be variously defined by prevailing management issues in the locality.

In these guidelines, the term 'coastal area' is preferred to 'coastal zone' to refer to the geographic entity covered by an integrated coastal management plan. Such coastal management takes place at two levels: the national, or subnational, level, where national goals, strategies, institutional arrangements and legislation may be determined and put into place; and the local, or area, level, where area-specific goals, objectives, plans and their implementation are the focus of attention.

The use of integrated coastal area management (ICAM) makes explicit the fact that degradation of coastal resources may result from activities outside the coastal zone, as defined at the beginning of these guidelines. Where issues are deemed to arise in a watershed, ICAM may, subject to the appropriate institutional arrangements, extend outside the coastal area.

The Economic and Environmental Importance of Coastal Areas

Favourable biophysical and climatic conditions, together with the ease of communication and navigation frequently offered by coastal sites (by sea or up river valleys), have encouraged human settlement in coastal zones since prehistoric times.

Economic Role

Many of the world's major cities are located in coastal areas, and a large portion of economic activities, with the exception of agriculture, are concentrated in these cities. The coastal zone is an area of convergence of activities in urban centres, such as shipping in major ports, and wastes generated from domestic sources and by major industrial facilities. Thus, traditional resource-based activities, such as coastal fisheries, aquaculture, forestry and agriculture, are found side by side with activities such as industry, shipping and tourism.

The potential for economic opportunities in coastal cities is a strong attractive force, fuelling immigration, often from economically depressed rural areas. As a result, in the future much larger, younger populations can be expected in the coastal areas of developing countries. These future coastal residents will demand employment, housing, energy, food, water and other goods and services, thus presenting a substantial development challenge.

Against this demographic backdrop, coastal areas are extremely important for the social and economic welfare of current and future generations, as coastal resources support key economic and subsistence activities. The economies of most developing countries are currently very dependent on natural resources, for agriculture, fisheries and forestry subsectors, mining, oil and gas extraction, marine tourism and ocean transport. Many of the world's most productive agricultural areas are located in river deltas and coastal plains. In particular, the deltas' food productivity exceeds local consumption needs and eventual delta disturbance can result in national economic shock waves that reach far beyond the delta.

Although, in the future, coastal areas will become more urbanized, and the economies of developing countries will undoubtedly diversify to some extent through industrialization, dependence on coastal resources is likely to remain strong. Industrial development often entails the processing of agricultural, fishery and forestry products, together with oil refining and textile manufacture. These diversified economic activities are often also dependent on coastal resources and, as economic diversification increases and makes the component sectors more interdependent, conflicts over natural resources and the environment will tend to develop.

Environmental Role

Coastal areas are also important ecologically, as they provide a number of environmental goods and services. The peculiar characteristic of coastal environments is their dynamic nature which results from the transfer of matter, energy and living organisms between land and sea systems, under the influence of primary driving forces that include short-term weather, long-term climate, secular changes in sea level and tides.

Marine, estuary and coastal wetland areas often benefit from flows of nutrients from the land and also from ocean upwelling which brings nutrient-rich water to the surface. They thus tend to have particularly high biological productivity. Moreover, coastal areas frequently contain critical terrestrial and aquatic habitats, particularly in the tropics. Such habitats together comprise unique coastal ecosystems, support a rich biological diversity and frequently contain a valuable assortment of natural resources. Examples of such habitats are estuarine areas, coral reefs, coastal mangrove forests and other wetlands, tidal flats and seagrass beds, which also provide essential nursery and feeding areas for many coastal and oceanic aquatic species.

It is estimated that 90 percent of the world's fish production is dependent on coastal areas at some time in their life cycle. In addition, these areas support large numbers of migratory and non-migratory waterfowl and shorebirds, and endangered reptiles, such as turtles and alligators. The advantages of maintaining their biological diversity have been formally recognized.

Physical features of coastal ecosystems, such as reefs and belts of mangrove, are important for the mitigation of the effects of natural disasters, such as storm-tide surges, shoreline retreat or floods. These features also play an essential role in natural processes, such as land accretion, and help to control coastal erosion and other damage arising from wind and wave action.

Even when coastal areas do not provide unique biological ecosystems, their location at the sea/land interface has recreational and aesthetic values which, in many countries, support valuable tourism activities, as well as providing attractive sites for industrial development and human settlements. The recreational and aesthetic values of coastal areas are increasing in developing countries as coastal tourism develops and domestic demand rises with increasing real incomes. Unique and appealing vistas, sandy and rocky beaches, pristine blue water, wetlands and coastal forest, and the associated wildlife, coral reefs and multiple recreational activities supported by these areas are major attractions of coastal areas.

Environmental Threats to Coastal Areas

The dynamics of alluvial landscapes and natural sedimentation patterns that determine the nutrient and energy flows in coastal areas are increasingly being modified by human activities, in particular those that affect water flows (dams, increased water extraction, deviation of rivers) and erosion, especially that caused by deforestation. This prevents or slows down vertical accretion, thus aggravating salt-water intrusion and impairing drainage conditions in riverine, delta or estuarine areas. It reduces or blocks sediment supply to the coast itself, which may give rise to the retreat of the coastline through wave erosion. Coastal areas are also prone to threats from natural causes such as tidal surges and sea-level rise.

The worst scenario projects a sea-level rise of 95 cm by the year 2100, with large local differences (resulting from tides, wind and atmospheric pressure patterns, changes in ocean circulation, vertical movements of continents, etc.) in the relative sea-level rises. The impacts of sea-level rise are therefore expected to be more local than global. The relative change of sea and land is the main factor. Many cities, for instance, suffer land subsidence as a result of groundwater withdrawal. This may be compounded by sea-level rise, especially since rates of subsidence may exceed the rate of sea-level rise between now and 2100. Under the worst scenario, the majority of the people who would be affected live in China (72 million), Bangladesh (13 million people and loss of 16 percent of national rice production) and Egypt (6 million people and 12 to 15 percent of agricultural land lost). Between 0.3 percent (Venezuela) and 100 percent (Kiribati and the Marshall Islands) of the population would be affected. Even more significant than the direct loss of land caused by the sea rising, are the associated indirect factors, including erosion patterns and damage to coastal infrastructure, salinization of wells, suboptimal functioning of the sewage system of coastal cities (with resulting health impacts), loss of littoral ecosystems and loss of biotic resources. In coastal areas, and particularly deltas, factors such as modified ocean circulation patterns (and their impact on building and erosion of the coast), climate change in the catchment basin and change in coastal climate, not to mention changes in the frequency of extreme events, should be taken into account.

While increased damage to coastal areas will certainly occur, it will be linked more to development (value of assets) than to increased fragility of populations or agriculture. Particularly in small islands, where development leads to a concentration of wealth, there is a risk of high damage being caused by natural disaster events. Where the extremes of one or more variables are involved at the same time, the risks will be very high. Such 'time bombs' are to be found in highly vulnerable deltas (which are densely populated and normally feed a large hinterland) that may become very unstable over a short period of time.

Wherever rising populations continue to be economically dependent on the primary producing sectors Đ agriculture, forestry and fisheries Đ and on other sectors directly dependent on natural resources, such as tourism and mining, the overexploitation of renewable and non-renewable resources, the degradation of ecosystems and the resulting loss of natural resource productivity are of primary concern. In this context, it is worth noting that marine pollution is derived mainly from land-based sources and the atmosphere, with river runoff and land-based discharges directly affecting coastal waters. More than 90 percent of all chemicals, refuse and other material entering coastal waters remains there in sediments, wetlands, fringing reefs and other coastal ecosystems.

Demographic and Occupational Pressures

According to the distribution of population in relation to the distance from the nearest coastline, 20.6 percent of the world's population lives within 30 km of the coast, and 37 percent within 100 km. As a result of migration to coastal areas, and in particular to coastal cities, the coastal population is growing at a faster rate than the world population; within the next 20 to 30 years, the coastal population is projected to almost double. At present, two-thirds of the world's cities with a population of 2.5 million or more are situated near tidal estuaries and 220 million people live in the, mostly coastal, megacities. These urban developments are taking up fertile agricultural land and leading to pollution of rivers, estuaries and seas by sewage and industrial and agricultural effluents. In turn, this is posing a threat to coastal ecosystems, their biological diversity, environmental regulatory functions and role in generating employment and food.

Poor people often migrate from inland rural areas and settle along the coast in search of better livelihoods because they cannot find employment elsewhere. They are drawn to coastal areas where resources still tend to be 'open access', or freely available, and thus offer sources of income of last resort. One means of relieving the pressures on coastal areas is therefore to improve living conditions in more distant rural areas. Investments in natural resource conservation in inland coastal areas, often made up of hilly and fragile land, will have the double advantage of improving conditions for poor farmers and reducing erosion and siltation. The creation of non-agricultural employment and the introduction of appropriate management systems will further improve the situation.

The World Resources Institute developed an index of development-related threats to coastal ecosystems, and defined coastal zones as including the land area within 60 km of adjacent near-shore waters. The index is based on five globally available georeferenced indicators of potential anthropogenic threats:

- cities with populations of more than 100 000: captures potential threats from coastal development, sewage, and industrial pollution;
- major ports: measures the potential threats from species introductions (through the release of ballast water), the potential for oil spills and industrial pollution;
- population density: measures potential threats from coastal development and pollution;
- road density: an indirect measure of access to coastal resources and coastal development;
- pipeline density: measures the potential threat of oil pollution and of spills of other industrial wastes.

It is important to note that, as the above indicators measure pressure rather than condition, the results do not imply that these areas have actually been degraded. In addition: the impacts of fishing, deforestation and agricultural activity are not covered; human activities beyond 60 km of the coast were not considered; the study did not include the relative sensitivities of different ecosystems to disturbance; data quality was better for some regions than for others; data modelling and mapping added additional uncertainty to the results; and pressures may have been under-represented where they have a cumulative effect. The current study was also at too coarse a scale to guide national management and planning activities and the intention is to develop a set of more comprehensive, finer-resolution indicators.

Sectoral Interdependencies in Coastal Areas

The coastal resource system is interdependent and interrelated and has direct and indirect connections with inland resource systems. For instance, production of fish may be dependent on the habitat for juveniles provided by mangrove swamps, and the health of a coral reef may be related to the filtering properties of the mangrove ensuring that only clear water reaches the reef. Conversely, a coral reef may die as a result of being covered with silt from soil erosion, perhaps occurring many miles upstream, caused by inappropriate forestry or agricultural practices. Mangroves, coastal dunes and reefs may protect coastal agriculture from erosion or storm surge, for example.

Ill-managed economic development in coastal areas is likely to create serious problems related to water pollution, degradation of critical habitats, depletion of natural resource stocks and other effects. When this occurs, the much publicized benefits of growth such as increased employment and rising nominal incomes will be undermined by increased costs in the areas of health, productivity and aesthetics.

Economic activities consist in changing resources or inputs into products or services. Thus, all activities directly or indirectly affect and interact with their environment (i.e. ecological, economic and social systems). For example, air and water temperatures determine the types of agriculture, forestry and fisheries that are possible and affect productivity, while market characteristics affect their viability.

Such interactions may be categorized as being synergistic, complementary, competitive or antagonistic. Integrated coastal area management aims at maximizing synergistic and complementary interactions and minimizing competitive and antagonistic ones.

Causes of Ecosystem Degradation and Natural Resource Loss in Coastal Areas

Ecosystem and natural resource damage in coastal areas, and hence reduced environmental goods and services, stem from increasing demands on resources (especially local demand arising from demographic pressure) and unsustainable management practices. The causes of ecosystem or natural resource damage in coastal areas can be analysed according to three interrelated 'failures': market failure (of which a related issue is property rights); policy or intervention failure; and information failure.

Market Failure and Property Rights

Market prices frequently fail to reflect the true cost of a good or a service and so send incorrect signals to the marketplace; sometimes, putting a price on a good may be impossible. Such situations represent market failure.

An example of incorrect market signals is the lack of internalizing of the cost of conversion of mangrove forest to agriculture or aquaculture because the costs of the conversion do not internalize externalities. Mangrove may be taken over by landless rural people seeking to make a living. But where users (e.g. companies) pay the owner for the (use of the) land, it has often been seen that in future years productivity will decline, there will be saline intrusion and/or storm damage will occur inland because of the absence of the protection that the mangrove provided. None of these negative factors will be reflected in the purchase price. The cost of destroying the mangrove may well have to be borne, not by whoever destroys it, but by someone else; in this case, market failure is even more pronounced.

Widespread destruction of mangroves can lead to siltation of estuaries and ports or eutrophication of coastal waters; such impacts, which originate outside the places they affect and are almost invariably negative, being referred to as negative 'externalities'.

The term 'property right' refers to a legally enforceable right that may arise from legislation or unwritten common law or customary law. A distinction should be made between property rights as such and *de facto* situations of access to a resource because it is not owned by anyone (for instance fish or wild animals prior to capture) or because the owner (who may be the state) does not exercise the right to exclude others from access to the resource. The issue of property rights arises most frequently when a good or service cannot be priced.

For any property regime, it is essential that an authority system (e.g. state authority or traditional leaders) can meet the expectations of rights holders. When the authority system breaks down, management of the natural resource fails, and the entire system changes (i.e. common property degenerates into open access). The authority mechanisms and capacity to enforce compliance ensure compliance with, and integrity of, the property regime.

Property and use rights are fundamental to the allocation of natural resources, as they determine who has the legal right to control access to the resource. In ICAM, the issue arises when ways of preventing environmental degradation or overexploitation (such as mangrove conversion or overfishing) are being sought. However, the response may vary according to circumstances. For example: the state may choose to enforce an existing ownership right over mangroves to prevent conversion; the distribution of property rights could be changed (e.g. use rights may be granted to a limited group of users, or state-owned resources could be sold); or legislation could be introduced to restrict the rights of private owners where these are causing harm.

One of the areas where open access is most prevalent in coastal zones is in fisheries. Very often, any citizen of the state has, in effect, free access to fisheries. The result in recent years is that most fisheries are fully or overexploited. A newcomer entering such a fishery may have a better vessel and fishing gear than current fishers and may, as a result, be able to fish profitably. The newcomer's increased wealth, however, will be made at the expense of the more poorly equipped fishers.

In recent years, the degradation of open access resources has led to an increased awareness of the value to society of these resources. Much attention is therefore being given to establishing or rebuilding institutions to manage access to threatened resources to help to protect society from negative resource externalities.

The existence of private or common property or exclusive use rights may lead to undesirable environmental and other impacts. For example, the holding of private property rights by entrepreneurs investing in shrimp cultivation in some coastal areas of South and Southeast Asia and of Latin America has, in some cases, resulted in destruction of wetland forests, water pollution and the marginalization and dislocation of resource-poor people. Where population pressure increases, through migration for instance, and pure survival is the issue, it will often be difficult to ensure compliance with various property regimes that endeavour to exclude one group or another.

In a perfect market situation (which only exists in textbooks) resources would be allocated to their correct societal use. In practice, resources usually have multiple uses valued differently by different potential users (forest dwellers need wood for heating and cooking while environmentalists defend trees *per se*; a logging company may see only the value of the exploitable trees, while local communities are aware of the forest's non-wood resources such as medicinal plants, wildlife and insects, and its role in absorbing CO_2 or producing oxygen). Excluding people from access to essential resources, especially when survival itself depends on that access, has a high social cost and, indeed, eventually becomes impossible.

Owing to the difficulty of accounting for the environmental goods and services of a given resource, especially in the long term, and the social values placed upon them by different user groups, the market will be a poor guide to management decisions. Appropriate management will depend on the circumstances and interactions of social, economic and biophysical institutions.

Policy Failure

Policy failure is the single most important cause of natural resource loss and ecosystem degradation. Policy failure is usually driven by the pursuit of short-sighted economic gain and lack of awareness of the long-term implications of non-intervention.

In most cases, policy failure means policies have not responded to actual or threatened degradation in the resource base or ecosystems, or have not taken account of local views. This leads to weakening of the institutions needed to cope with the situation; absence of a legal framework or weak enforcement capabilities for example. Such situations may apply to coastal resources which therefore become open to overexploitation.

Resource loss and ecosystem degradation will only be halted if appropriate policies are adopted and implemented. Growing awareness of the environmental, social and economic consequences of policy failure that are of global concern, such as global warming or loss of biological diversity, is fortunately beginning to put pressures on governments to rectify present polices and create an appropriate policy environment.

More Specifically, the Following Measures should be Adopted

- legislation related to the long-term sustainability of the natural environment;
- decentralized structures established on the basis of effective participation at local level;
- incentives for producers;
- provision of support that cannot be covered from local resources.

In the current context of globalization and interdependence, international institutions often impose policy reforms related to market forces, which do not always take account of the needs of nations whose bargaining power is weak.

Global environmental concerns and the international exchange of goods and services are, however, governed by conventions and agreements that influence national and local decision-making. While economic efficiency and competitiveness are clearly important, local-level concerns ought to be voiced in the macropolicy arena in order to reflect and respect environmental, cultural and economic diversity.

Sustainability (or unsustainability) can be measured by the sum of the various human, economic and natural resources, where the degree of use, exchange and trade among them will vary according to the values given to each. The interactions between resources, and the degree of substitutability, therefore ultimately depend on the acceptability of trade-offs between resources.

In farming systems, trade-offs exist between productivity, stability, resilience and equity. The different properties of a system and their importance to different groups will determine the trade-off patterns. As farm systems evolve in response to their own logic, as well as to changes in society, agricultural innovations may initially be rejected because of socio-cultural constraints to their adoption, but will be adopted rapidly if economic circumstances change (e.g. capital, labour, markets, support services). In other words, the sustainability of a system depends on circumstances, that is, on contingency.

Information Failure

Sound policies require sound information. Frequently in developing countries, information is lacking, especially in such areas as renewable resources and, more particularly, on the status of renewable resources (especially fisheries), natural resource dynamics, land-use and tenure patterns, institutional, social and cultural conditions, and levels of investment in coastal areas. Similarly, environmental monitoring, for example of water quality, is often scant.

Perfect information will never be available even in the most advanced countries; for example, it is often extremely difficult to determine if a decline in fish catches in a particular area is caused by the effect of sedimentation of habitat, overfishing, natural factors or all three. However, the more information is available, the better the basis for policy decisions.

Lack of information can be a contributory factor to policy failure. For example, policy-makers may take the view that their country is too poor to afford conservation of renewable resources or ecosystems, and would opt for economic development even if it destroys some of the country's natural capital. The attitude may be different if policy-makers are aware of the full economic value of the resource under threat and take account of the value of all the goods and environmental services it provides.

The routine collection and analysis of information is costly, in both human and financial terms. But, insofar as policy decisions cannot be taken without adequate information, it must be made available. The cost of gathering information can be contained by identifying as precisely as possible the information requirements related to coastal areas that are under threat.

However, lack of adequate information should not be a reason for inaction. Useful action can be taken even where only relatively superficial information is available and there are often good arguments for taking a precautionary approach: 'when in doubt, protect'. In order to overcome these difficulties gradually, research requirements should be built into the strategy or plan, encompassing information gathering and strengthening staff capacities to analyse and compute various types of information and to analyse interactions.

Policy-makers need tools to assist them in policy-making and planning, especially to gauge whether their actions are likely to result in sustainable development. Some attempts to create indicators of ecological sustainability and quality of life have been made by a number of institutions. Such indicators, however, must be used with caution as situations are always highly complex and the quantification of the environment Ð and more especially its monetization Ð is fraught with difficulties.

Once policy-makers, planners and resource users agree on the issues, criteria for thresholds of resource degradation can be established. Factors here will include social and economic considerations, such as the time, expense and level of detail that is being aimed at. In addition, estimates and common sense based on experience, training and intuition will also often provide a satisfactory basis for decisions.

Agriculture, Forestry and Fisheries in Coastal Areas

Most developing countries are characterized by a fast-growing population and workforce. The rate of growth of the world's economically active population, between 1990 and 2000, is 1.86 percent per year. While nearly half of this workforce is absorbed in the agriculture, forestry and fisheries subsectors, an increasing amount of labour will seek employment in non-farm occupations. Finding productive and remunerative employment for these workers will be one of the main challenges of the twenty-first century.

The development of non-farm jobs in villages and small towns closely linked to agriculture and farm-produce processing is a powerful way of relieving pressure on coastal areas. It is especially efficient as the cost of job creation in rural areas is much less than in urban areas. It has the added advantages of:

- creating or maintaining employment related to the primary sector in coastal areas (through both agriculture and small agro-industry activities);
- stabilizing rural employment and communities;
- contributing to the overall distribution of wealth;
- reducing environmental damage;
- promoting traditional and informal activities;
- promoting links between agriculture and industry;
- reducing the cost of service activities.

However, while non-farm rural job creation will relieve pressures, the population of coastal areas will inevitably continue to increase. Worldwide, the ratio of economically active population in agriculture is declining.

The main consequences of mismanagement in coastal areas in general are felt in health, productivity and aesthetics. Some of these issues are intrasectoral, such as depletive/destructive resource uses. Others are often trans-sectoral, such as loss of habitat and environmental degradation.

Agriculture

In some coastal areas, and especially in small islands, agricultural production makes an extremely important contribution to the local economy or to national agricultural production. In countries such as Egypt and Bangladesh, the river deltas, with their fertile alluvial soils, play a major role in the agriculture sector.

Whatever the situation, there are a number of reasons for giving agriculture particular attention in integrated coastal resource management planning. Among these are the following:

- agriculture has major positive, but also potentially negative, effects on the coastal environment. Sustainable agricultural policies are therefore needed to minimize the negative impacts of inland agriculture on coastal areas;
- agriculture is usually concerned with the production of food, and thus has strategic and political significance for food and livelihood security. Often, however, in coastal areas, urban growth, the high value of urban goods and services and the political importance of urban centres marginalize coastal agriculture and attract labour away from primary-sector activities. Protecting and maintaining peri-urban agriculture in coastal areas can make a substantial contribution to providing food to cities (particularly large ones) and creating employment;
- agriculture may also provide raw materials to industry located in coastal areas and may therefore have considerable economic significance. In particular, fostering the linkage between agriculture and tourism offers opportunities in both sectors for demand and supply of food, beverage and employment.

Planning for coastal agricultural activities must make explicit allowance for the wide range of farmers' interests and activities, including non-farm activities, the limited flexibility that farmers have in production decisions, their high vulnerability to adverse environmental change and the importance of non-commercial considerations in their decision-making. The complexity of these factors makes participation or consultation particularly important in coastal area planning where it impinges on agricultural activities. In particular, planning in coastal areas must account for such special characteristics of agriculture, such as the following:

- it includes a broad range of activities with a variety of types of product (crops and livestock, food and industrial raw materials), enterprises (from smallholdings to large commercial enterprises) and production processes (intensive, extensive);
- it makes a variety of contributions to the local economy, social cohesion and the maintenance of the cultural traditions of a society;
- it is dependent on natural growth processes and, to a greater or lesser extent, on land and its associated natural resources and environment.

Small-scale agriculture in many developing countries deploys strategies that are governed by complex considerations among which economic factors may rank very low. Last but not least, it is essential to take the special requirements of gender fully into account.

Forestry

Coastal forests in developing countries are mainly mangroves, which cover an estimated 15.5 million hectares worldwide. Other coastal forested ecosystems include savannah woodlands, dry forests and rain forests. Commercial production from mangroves is comprised of building materials, such as poles and timber, and numerous non-forest products. Savannah woodlands and dry forests are used primarily for grazing, while rain forests are used for their commercial timber.

The value of mangroves is primarily environmental and social, rather than economic. An important part of the value of mangroves lies in the environmental services provided by their unique habitat which is host to significant biological diversity and provides spawning and nursery grounds for many species of commercially valuable fish. Mangroves also act as a sediment trap, a source of nutrients to inshore waters, and help to protect areas against shoreline erosion and surge-tide damage.

Economic activities associated with forests, even inland, may be critical to the environmental health of the coastal area. In particular, excessive cutting and extraction of timber on steep slopes will result in severe erosion, which is likely to lead to turbidity in rivers carrying away the displaced topsoil, and sedimentation of fish habitats, such as seagrass beds and coral reefs, degrading and ultimately destroying them. In Bacuit Bay, the Philippines, for example, the benefits of logging activities on mountain slopes are outweighed by the loss in tourism and fishing value resulting from the degradation of offshore reefs.

Coastal forests are often perceived as an obstacle to development, rather than a resource to be supported. As absolute priority is often given to the immediate and most pressing economic needs, medium- and long-term gains may be overlooked.

Fisheries

The fisheries sector often accounts for a significant proportion of GDP, up to 5-10 percent in the economies of many developing coastal nations. Revenue earned by some small island developing states from the exploitation and processing of their fisheries resources accounts annually for more than 50 percent of public-sector revenue. The sustainable development of coastal fisheries (which, for example, represent less than 10 percent of total catch in the Pacific but, significantly, 70 percent of local consumption) will determine the ability of many coastal communities to survive. A large proportion of the total fish production of many coastal countries is usually derived from coastal fisheries. At the local level, the fisheries sector provides a livelihood to the population and also contributes to food security. Annual average per caput consumption of fish is 9 kg in developing countries, 27 kg in developed countries and 50 kg in many small island states (most of the population's animal protein intake). Coastal areas everywhere provide habitat for about 90 percent of marine fish in commercial and subsistence fisheries, at all or some stages in the lives of the fish.

Conventional fisheries management has been concerned only with problems generated within the fisheries sector. Planning in coastal areas has to take account of the specific features of small-scale and artisanal fisheries, including the risk of overexploitation by 'last resort' fishers (poor people with no other source of livelihood). Solutions include the creation of jobs in other coastal sectors or support to help fishers to fish further offshore thus relieving pressures on inshore areas. Increasingly, policy reforms aiming at the conservation of coastal resources and habitats take the form of revitalizing traditional fisheries management systems with a view to incorporating them into management planning.

Coastal aquaculture has been growing in a number of industrialized and developing countries, particularly in parts of Southeast and East Asia and parts of Latin America, and there is considerable scope for development and expansion in other regions. However, while coastal aquaculture is gaining economic importance in many countries, it can also have damaging consequences if it is poorly managed. Moreover, it is susceptible to environmental change originating in actions outside the sector. Coastal aquaculture must be environmentally managed and absorbed into the coastal management planning process.

Coastal Development Hazards

A coastal development hazard is something that affects the natural environment by man-made products.

As coasts become more developed, the vulnerability component of the equation increases as there is more value at risk to the hazard. The likelihood component of the equation also increases in terms of there being more value on the coast so a higher chance of hazardous situation occurring. Fundamentally humans create hazards with their presence. In a coastal example, erosion is a process that happens naturally on the Canterbury Bight as a part of the coastal geomorphology of the area and strong long shore currents. This process becomes a hazard when humans interact with that coastal environment by developing it and creating value in that area.

In Burton 1978 *'The Environment as Hazard'* a natural hazard is defined as the release of energy or materials that threaten humans or what they value. In a coastal context these hazards vary temporally and spatially from a rare, sudden, massive release of energy and materials such as a major storm event or tsunami, to the continual chronic release of energy and materials such long-term coastal erosion or sea-level rise.

Coastal Population Growth and Development on Coasts

Globally the number of people living on the coast is increasing. It has been stated that there has been over a 35% increase in the population of people living on the coasts since 1995. The average density of people in coastal regions is 3 times higher than the global average density. Historically city development especially large cities were based on coasts due to the economic benefits of the ports. In 1950 there were only 2 megacities (cities with greater than 8 million people) in the coastal zone, London and New York. By the mid-nineties there were 13. Although coastal areas globally have shown population growth and increases in density, very few in-depth quantitative global studies of population have been carried out, especially in terms of distribution across specific environs, like coasts. The spatial distribution and accuracy of global data must be significantly improved before realistic quantitative assessments of the global impacts of coastal hazards can be made. As currently much of the data is collected and analysed in the aftermath of disasters.

Historic studies have put estimates of the number of deaths due to cyclones over the last 200 years around the Bay of Bengal exceeding 1.3 million. However in developed countries, as can be expected, the death toll is significantly lower but the economic losses due to coastal hazards are increasing. The United States of America (USA) for example had major losses in Hurricane Andrew, which hit Florida and Louisiana in 1992.

This rushing to the coast is exhibited in property value. A study by Bourassa et al. (2004) found in Auckland New Zealand, wide sea views contributed on average an additional 59% to the value of a waterfront property. This effect diminished rapidly the further from the property was from the coast. In another study it was found that moving 150m away from the Gulf of Mexico lowered property values by 36%.

Insurance premiums in coastal hazard areas are an inconsequential determinant of property values, given the significant amenity values provided by the coast in terms of views and local recreation. Sea-level rise, coastal erosion, and the exacerbating interaction between these two natural phenomena are likely to pose a significant threat for the loss of capital assets in coastal areas in the future. It is hard to say if the vulnerability to coastal hazards by those residing there is perceived yet dominated by the amenity value of coasts or simply ignored.

Coastal Erosion Hazards

Coastal erosion is one of the most significant hazards associated with the coast. Not in terms of a rare massive release of energy or material resulting in loss of life, as is associated with tsunami and cyclones, but in terms of a continual chronic release that forms a threat to infrastructure, capital assets and property.

Beach Erosion Process

Storm induced large erosion events are a part of the natural evolutionary process of fine sediment, gently sloping beaches. Increased wave energy in storms leads to the removal of foreshore, berm and dune sediments. These displaced sediments are then deposited as near shore sand bars and act to dampen the wave energy lessoning the amount of sediment that is being eroded from the coast. When wave energies decrease post storm events, the sediments from these newly deposited near shore bars are returned to the upper beach, rebuilding the berm. This self-correcting cycle is an active balance between wave energies and fine sediment deposition. This store of sediment being available for erosion in storms and re-depositing when the event has subsided is an important natural buffer mechanism against protecting the mainland from erosion and minimising coastal retreat.

Dune Destruction

Sand dunes are very dynamic fragile structures that act as stores of sediment used to carry out the coastal processes mentioned above. This removal of the upper beach sediments is important from a hazard perspective as this is the area of the coast that is often utilised for property development due to the high prices sea front properties with a view can achieve. In Pegasus Bay, New Zealand, storm events in 1978 and 2001 caused significant erosion of the New Brighton and Waimairi sand beaches. In the 1978 storm event houses on the seaward side of the New Brighton Spit suffered from undercutting as the dune sediment in which they were built on was eroded by high wave energy. This same storm event caused similar erosion damage to houses built on the upper dunes in Raumati, on the west coast of the North Island, New Zealand. Bulldozing and bulk removal of sand from protective coastal dunes is therefore an extremely hazardous activity, and one that has been widely carried out in New Zealand in order to form a surface on which to build on to obtain sea views.

Canterbury Bight

In Kirk (2001) coastal erosion on the Canterbury bight, South Canterbury was said to have reached up to 8 m per year. This coastal process could be measured in more ways than one, the afore mentioned distance of coastal retreat or the decreased dollar value of developed assets, land and infrastructure that are at risk. To date, erosion on the Canterbury Bight has led to the loss of agricultural land, threatened valuable infrastructure including holiday settlements, and reduced coastal lagoons and wetlands.

Historically, erosion on the Canterbury Bight was a natural process, but has now been exacerbated by human intervention. The Waitaki River was the dominant source of sediments for the beaches between Oamaru and Timaru. Since the damming of the Waitaki River in 1935 erosion of the coastal cliffs has become the primary source of sediment in the north flowing current moving up the coast of South Canterbury.

Erosion Mitigation

Engineered Structures

This destruction of sand dunes is often then mitigated with construction of seawalls, revetments and groynes in often futile attempts to prevent storm erosion hazards to unsuitably located assets and infrastructure on coasts. These engineered methods are commonly ineffective and frequently actually magnify the hazard or just move the hazard down coast. In Porthcawl, South Wales, a seawall constructed to stop erosion in 1887 was replaced in 1906, 1934 and finally in 1984 when the beach was paved as each prior structure was undermined by further erosion. The loss of aesthetics due to the lack of a sand beach resulted in tourists utilising alternative beaches. Therefore incurring an even greater economic loss on top of the cost of the engineering.

Restorative Dune Planting

The alternative to hard engineering measures is sand dune conservation. This involves protecting the sand dunes and allowing the natural buffering processes to occur. Dune protection and conservation can be facilitated in a number of ways, actively with dune planting and sand fencing, or with better planning by developing away from or well behind the dune structures not on them. On the New Brighton Spit the spread of marram grass *(Ammophila arenaria)* has resulted in effective dune stabilisation in areas. However this invasive exotic species has mostly replaced indigenous species like pingao *(Desmoschoenus spiralis)* meaning that although the stability of the coastal area has gained, the historic, native cultural values of the area have suffered.

Beach Nourishment

A further soft-engineering method for protecting the shoreline is beach nourishment, due to cost this is a solution that has been used primarily for the benefit of the tourism industry. As a result of erosion Miami Beach had almost no stored sediment left by the mid 1970s, consequently, visitor numbers declined and development of the area decreased. A beach nourishment program was set up resulting in an influx of development and infrastructure in the late 1970s. Miami Beach was rejuvenated to such an extent that annual revenue from foreign tourists alone is \$2.4 billion, compared to the \$52 million cost of the 20-year nourishment project. Tax revenue from tourists who visit Miami Beach alone more than covers the cost of beach nourishment projects across the nation. Using the capitalised annual cost of the project, for every \$1 that has been invested annually on the nourishment, Miami Beach has received almost \$500 annually in foreign exchange.

Sea-Level Changes

Sea level is generally used to refer to mean sea level (MSL), an average level for the surface of one or more of Earth's oceans from which heights such as elevations may be measured. MSL is a type of vertical datum – a standardised geodetic reference point – that is used, for example, as a chart datum in cartography and marine navigation, or, in aviation, as the standard sea level at which atmospheric pressure is measured in order to calibrate altitude and, consequently, aircraft flight levels. A common and relatively straightforward mean sea-level standard is the midpoint between a mean low and mean high tide at a particular location.

Figure: *This marker indicating sea level is situated between Jerusalem and the Dead Sea, Israel.*

Sea levels can be affected by many factors and are known to have varied greatly over geological time scales. The careful measurement of variations in mean sea levels can offer information about climate change and has been interpreted as evidence supporting the view that the current rise in sea levels is an indicator of global warming.

The term above sea level generally refers actually to above mean sea level (AMSL).

Measurement

Precise determination of a "mean sea level" is a difficult problem because of the many factors that affect sea level. Sea level varies quite a lot on several scales of time and distance. This is because the sea is in constant motion, affected by the tides, wind, atmospheric pressure, local gravitational differences, temperature, salinity and so forth. The best one can do is to pick a spot and calculate the mean sea level at that point and use it as a datum. For example, a period of 19 years of hourly level observations may be averaged and used to determine the mean sea level at some measurement point.

To an operator of a tide gauge, MSL means the "still water level"—the level of the sea with motions such as wind waves averaged out—averaged over a period of time such that changes in sea level, e.g., due to the tides, also get averaged out. One measures the values of MSL in respect to the land. Hence a change in MSL can result from a real change in sea level, or from a change in the height of the land on which the tide gauge operates.

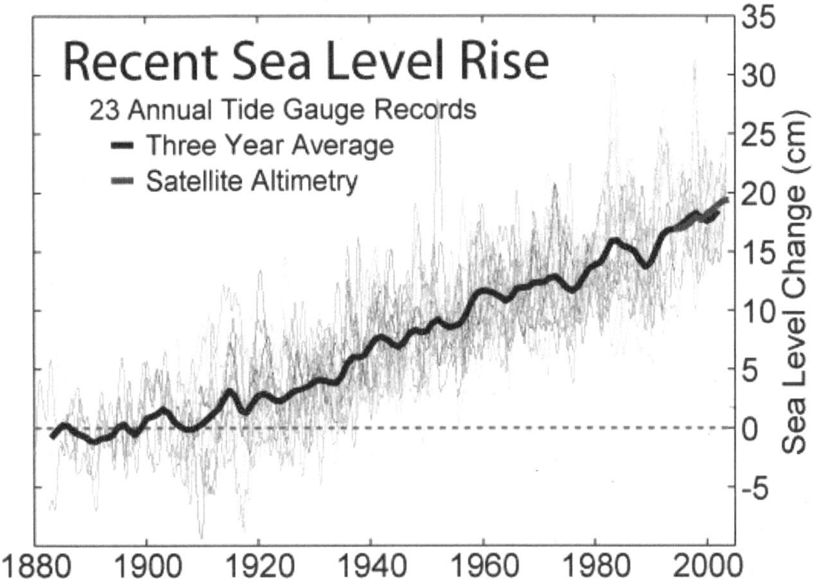

Figure: *Sea level measurements from 23 long tide gauge records in geologically stable environments show a rise of around 200 millimetres (7.9 in) during the 20th century (2 mm/year).*

In the UK, the Ordnance Datum (the 0 metres height on UK maps) is the mean sea level measured at Newlyn in Cornwall between 1915 and 1921. Prior to 1921, the datum was MSL at the Victoria Dock, Liverpool.

In France, the Marégraphe in Marseilles measures continuously the sea level since 1883 and offers the longest collapsed data about the sea level. It is used for a part of continental Europe and main part of Africa as official sea level. Elsewhere in Europe vertical elevation references (European Vertical Reference System) are made to the Amsterdam Pile elevation, which dates back to the 1690s.

Satellite altimeters have been making precise measurements of sea level since the launch of TOPEX/Poseidon in 1992. A joint mission of NASA and CNES, TOPEX/Poseidon was followed by Jason-1 in 2001 and the Ocean Surface Topography Mission on the Jason-2 satellite in 2008.

Height above Mean Sea Level

Height above mean sea level (*AMSL*) is the elevation (on the ground) or altitude (in the air) of an object, relative to the average sea level datum. AMSL height is used extensively in radio (both in broadcasting and other telecommunications uses) to determine the coverage area a station will be able to reach. It is also used in aviation, where some heights are recorded and reported with respect to mean sea level (MSL) (contrast with flight level), and in the atmospheric sciences, and land surveying. An alternative is to base height measurements on an ellipsoid of the entire earth, which is what systems such as GPS do. In aviation, the ellipsoid known as World Geodetic System 84 is increasingly used to define heights, however, differences up to 100 metres (328 feet) exist between this ellipsoid height and mean tidal height. The alternative is to use a geoid based vertical datum such as NAVD88.

When referring to geographic features such as mountains on a topographic map, variations in elevation are shown by contour lines. The elevation of a mountain denotes the highest point or summit and is typically illustrated as a small circle on a topo map with the AMSL height shown in either metres or feet or both.

In the rare case that a location is below sea level, the elevation AMSL is negative. For one such case see Amsterdam Airport Schiphol.

Difficulties in Utilization

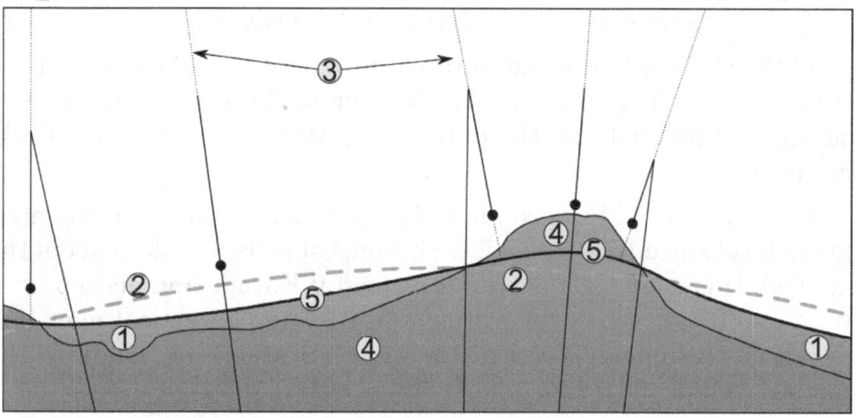

1. Ocean. 2. Reference ellipsoid.

3. Local plumb line. 4. Continent. 5. Geoid

To extend this definition far from the sea means comparing the local height of the mean sea surface with a "level" reference surface, or datum, called the geoid. In a state of rest or absence of external forces, the mean sea level would coincide with this geoid surface, being an equipotential surface of the Earth's gravitational field. In reality, due to currents, air pressure variations, temperature and salinity variations, etc., this does not occur, not even as a long term average. The location-dependent, but persistent in time, separation between mean sea level and the geoid is referred to as (stationary) ocean surface topography. It varies globally in a range of ± 2 m.

Historically, adjustments were made to sea-level measurements to take into account the effects of the 235 lunar month Metonic cycle and the 223-month eclipse cycle on the tides.

Sea Level and Dry Land

Figure: *Sea Level sign seen on cliff (circled in red) at Badwater Basin, Death Valley National Park.*

Several terms are used to describe the changing relationships between sea level and dry land. When the term "relative" is used, it means change relative to a fixed point in the sediment pile. The term "eustatic" refers to global changes in sea level relative to a fixed point, such as the centre of the earth, for example as a result of melting ice-caps. The term "steric" refers to global changes in sea level due to thermal expansion and salinity variations. The term "isostatic" refers to changes in the level of the land relative to a fixed point in the earth, possibly due to thermal buoyancy or tectonic effects; it implies no change in the volume of water in the oceans. The melting of glaciers at the end of ice ages is one example of eustatic sea level rise. The subsidence of land due to the withdrawal of groundwater is an isostatic cause of relative sea level rise. Paleoclimatologists can track sea level by examining the rocks deposited along coasts that are very tectonically stable, like the east coast of North America. Areas like volcanic islands are experiencing relative sea level rise as a result of isostatic cooling of the rock which causes the land to sink.

On other planets that lack a liquid ocean, planetologists can calculate a "mean altitude" by averaging the heights of all points on the surface. This altitude, sometimes referred to as a "sea level", serves equivalently as a reference for the height of planetary features.

Sea Level Change

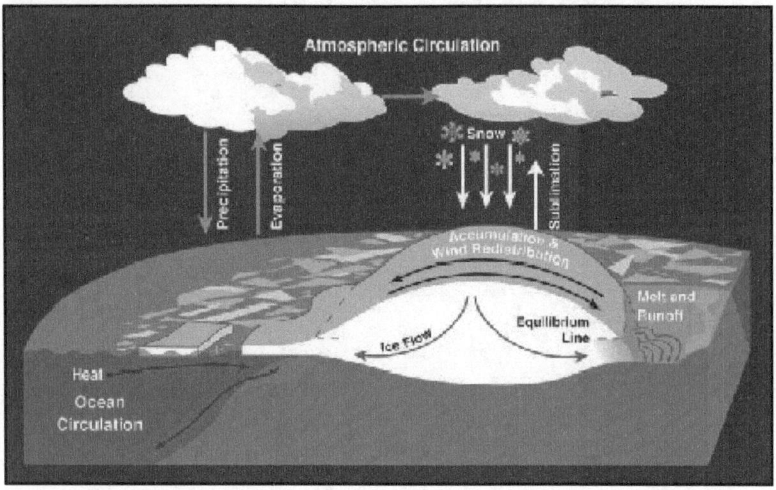

Water cycles between ocean, atmosphere and glaciers.

Local and Eustatic Sea Level: Local mean sea level (LMSL) is defined as the height of the sea with respect to a land benchmark, averaged over a period of time (such as a month or a year) long enough that fluctuations caused by waves and tides are smoothed out. One must adjust perceived changes in LMSL to account for vertical movements of the land, which can be of the same order (mm/yr) as sea level changes. Some land movements occur because of isostatic adjustment of the mantle to the melting of ice sheets at the end of the last ice age. The weight of the ice sheet depresses the underlying land, and when the ice melts away the land slowly rebounds. Changes in ground-based ice volume also affect local and regional sea levels by the readjustment of the geoid and true polar wander. Atmospheric pressure, ocean currents and local ocean temperature changes can affect LMSL as well.

Eustatic change (as opposed to local change) results in an alteration to the global sea levels due to changes in either the volume of water in the world oceans or net changes in the volume of the ocean basins.

Short Term and Periodic Changes: There are many factors which can produce short-term (a few minutes to 14 months) changes in sea level.

	Periodic sea level changes	
Diurnal and semidiurnal astronomical tides	12–24 h P	0.2–10+ m
Long-period tides		
Rotational variations (Chandler wobble)	14 month P	
	Meteorological and oceanographic fluctuations	
Atmospheric pressure	Hours to months	-0.7 to 1.3 m
Winds (storm surges)	1–5 days	Up to 5 m
Evaporation and precipitation (may also follow long-term pattern)	Days to weeks	
Ocean surface topography (changes in water density and currents)	Days to weeks	Up to 1 m
El Niño/southern oscillation	6 mo every 5–10 yr	Up to 0.6 m
	Seasonal variations	
Seasonal water balance among oceans (Atlantic, Pacific, Indian)		
Seasonal variations in slope of water surface		
River runoff/floods	2 months	1 m
Seasonal water density changes (temperature and salinity)	6 months	0.2 m
	Seiches	
Seiches (standing waves)	Minutes to hours	Up to 2 m
	Earthquakes	
Tsunamis (generate catastrophic long-period waves)	Hours	Up to 10 m
Abrupt change in land level	*Minutes*	*Up to 10 m*

Long Term Changes: Various factors affect the volume or mass of the ocean, leading to long-term changes in eustatic sea level. The primary influence is that of temperature on seawater density and the amounts of water retained in rivers, aquifers, lakes, glaciers, polar ice caps and sea ice. Over much longer geological timescales, changes in the shape of the oceanic basins and in land/sea distribution will also affect sea level. Observational and modelling studies of mass loss from glaciers and ice caps indicate a contribution to sea-level rise of 0.2 to 0.4 mm/yr averaged over the 20th century. Over this last million years, whereas it was higher most of the time before then, sea level was lower than today.

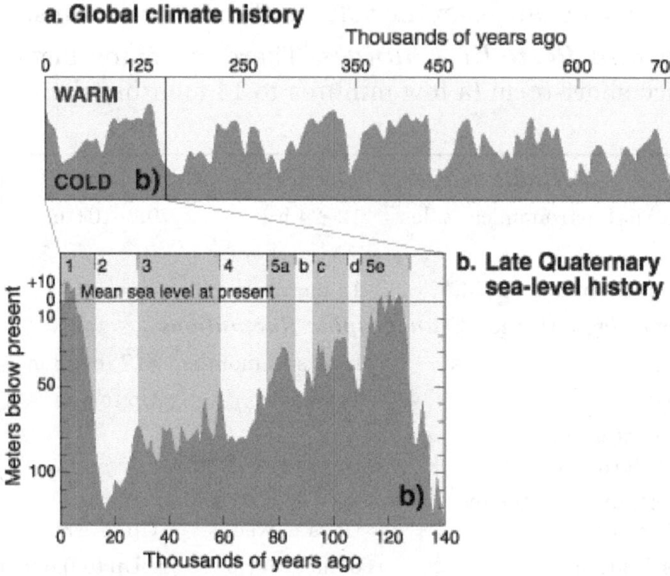

Figure: *Sea-level changes and relative temperatures*

Sea level reached 120 meters below current sea level at the Last Glacial Maximum 19,000–20,000 years ago.

Glaciers and Ice Caps: Each year about 8 mm (0.3 inches) of water from the entire surface of the oceans falls onto the Antarctica and Greenland ice sheets as snowfall. If no ice returned to the oceans, sea level would drop 8 mm (0.3 in) every year. To a first approximation, the same amount of water appeared to return to the ocean in icebergs and from ice melting at the edges. Scientists previously had estimated which is greater, ice going in or coming out, called the mass balance, important because it causes changes in global sea level. High-precision gravimetry from satellites in low-noise flight has since determined that in 2006, the Greenland and Antarctic ice sheets experienced a combined mass loss of 475 ± 158 Gt/yr, equivalent to 1.3 ± 0.4 mm/yr sea level rise. Notably, the acceleration in ice sheet loss from 1988–2006 was 21.9 ± 1 Gt/yr² for Greenland and 14.5 ± 2 Gt/yr² for Antarctica, for a combined total of 36.3 ± 2 Gt/yr². This acceleration is 3 times larger than for mountain glaciers and ice caps (12 ± 6 Gt/yr²).

Ice shelves float on the surface of the sea and, if they melt, to first order they do not change sea level. Likewise, the melting of the northern polar ice cap which is composed of floating pack ice would not significantly contribute to rising sea levels.

However, because floating ice pack is lower in salinity than seawater, their melting would cause a very small increase in sea levels, so small that it is generally neglected.

Scientists previously lacked knowledge of changes in terrestrial storage of water. Surveying of water retention by soil absorption and by artificial reservoirs ("impoundment") show that a total of about 10,800 cubic kilometres (2,591 cubic miles) of water (just under the size of Lake Huron) has been impounded on land to date. Such impoundment masked about 30 mm (1.2 in) of sea level rise in that time.

Conversely estimates of excess global groundwater extraction during 1900–2008 totals ?4,500 km3, equivalent to a sea-level rise of 12.6 mm (0.50 in) (>6% of the total). Furthermore, the rate of groundwater depletion has increased markedly since about 1950, with maximum rates occurring during the most recent period (2000–2008), when it averaged ~145 km3/yr (equivalent to 0.40 mm/yr of sea-level rise, or 13% of the reported rate of 3.1 mm/yr during this recent period).

If small glaciers and polar ice caps on the margins of Greenland and the Antarctic Peninsula melt, the projected rise in sea level will be around 0.5 m (1 ft 7.7 in). Melting of the Greenland ice sheet would produce 7.2 m (23.6 ft) of sea-level rise, and melting of the Antarctic ice sheet would produce 61.1 m (200.5 ft) of sea level rise. The collapse of the grounded interior reservoir of the West Antarctic Ice Sheet would raise sea level by 5 m (16.4 ft) - 6 m (19.7 ft).

The snowline altitude is the altitude of the lowest elevation interval in which minimum annual snow cover exceeds 50%. This ranges from about 5,500 metres (18,045 feet) above sea-level at the equator down to sea level at about 70° N&S latitude, depending on regional temperature amelioration effects. Permafrost then appears at sea level and extends deeper below sea level polewards.

As most of the Greenland and Antarctic ice sheets lie above the snowline and/or base of the permafrost zone, they cannot melt in a timeframe much less than several millennia; therefore it is likely that they will not, through melting, contribute significantly to sea level rise in the coming century. They can, however, do so through acceleration in flow and enhanced iceberg calving.

Climate changes during the 20th century are estimated from modelling studies to have led to contributions of between -0.2 and 0.0 mm/yr from Antarctica (the results of increasing precipitation) and 0.0 to 0.1 mm/yr from Greenland (from changes in both precipitation and runoff).

Estimates suggest that Greenland and Antarctica have contributed 0.0 to 0.5 mm/yr over the 20th century as a result of long-term adjustment to the end of the last ice age.

The current rise in sea level observed from tide gauges, of about 1.8 mm/yr, is within the estimate range from the combination of factors above but active research continues in this field. The terrestrial storage term, thought to be highly uncertain, is no longer positive, and shown to be quite large.

Geological Influences: At times during Earth's long history, the configuration of the continents and sea floor have changed due to plate tectonics. This affects global sea level by altering the depths of various ocean basins and also by altering glacier distribution with resulting changes in glacial-interglacial cycles. Changes in glacial-interglacial cycles are at least partially affected by changes glacier distributions across the Earth.

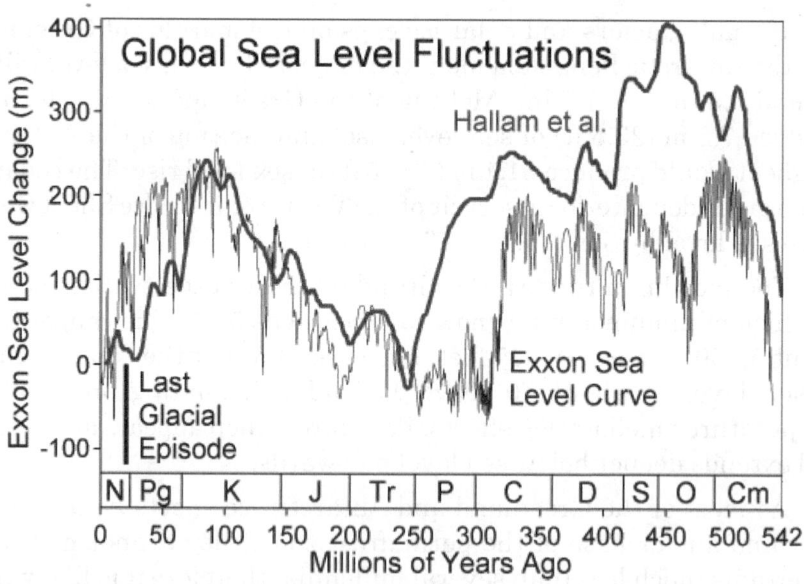

Figure: *Comparison of two sea level reconstructions during the last 500 Ma. The scale of change during the last glacial/interglacial transition is indicated with a black bar. Note that over most of geologic history, long-term average sea level has been significantly higher than today.*

The depth of the ocean basins is a function of the age of oceanic lithosphere (the tectonic plates beneath the floors of the world's oceans). As older plates age, they becomes denser and sink, allowing newer plates to rise and take their place.

Therefore, a configuration with many small oceanic plates that rapidly recycle the oceanic lithosphere would produce shallower ocean basins and (all other things being equal) higher sea levels. A configuration with fewer plates and more cold, dense oceanic lithosphere, on the other hand, would result in deeper ocean basins and lower sea levels.

When there was much continental crust near the poles, the rock record shows unusually low sea levels during ice ages, because there was much polar land mass on which snow and ice could accumulate. During times when the land masses clustered around the equator, ice ages had much less effect on sea level.

Over most of geologic time, the long-term mean sea level has been higher than today (see graph above). Only at the Permian-Triassic boundary ~250 million years ago was the long-term mean sea level lower than today.

Long term changes in the mean sea level are the result of changes in the oceanic crust, with a downward trend expected to continue in the very long term.

During the glacial-interglacial cycles over the past few million years, the mean sea level has varied by somewhat more than a hundred metres. This is primarily due to the growth and decay of ice sheets (mostly in the northern hemisphere) with water evaporated from the sea.

The Mediterranean Basin's gradual growth as the Neotethys basin, begun in the Jurassic, did not suddenly affect ocean levels. While the Mediterranean was forming during the past 100 million years, the average ocean level was generally 200 metres above current levels.

However, the largest known example of marine flooding was when the Atlantic breached the Strait of Gibraltar at the end of the Messinian Salinity Crisis about 5.2 million years ago.

This restored Mediterranean sea levels at the sudden end of the period when that basin had dried up, apparently due to geologic forces in the area of the Strait.

Long-term causes	***Range of effect***	***Vertical effect***
Change in volume of ocean basins		
Plate tectonics and seafloor spreading (plate divergence/convergence) and change in seafloor elevation (mid-ocean volcanism)	Eustatic	0.01 mm/yr
Marine sedimentation	Eustatic	< 0.01 mm/yr
Change in mass of ocean water		
Melting or accumulation of continental ice	Eustatic	10 mm/yr
• Climate changes during the 20th century		
• Antarctica	Eustatic	0.39 to 0.79 mm/yr
• Greenland (from changes in both precipitation and runoff)	Eustatic	0.0 to 0.1 mm/yr
• Long-term adjustment to the end of the last ice age		
• Greenland and Antarctica contribution over 20th century	Eustatic	0.0 to 0.5 mm/yr
Release of water from earth's interior	Eustatic	
Release or accumulation of continental hydrologic reservoirs	Eustatic	
Uplift or subsidence of Earth's surface (Isostasy)		
Thermal-isostasy (temperature/density changes in earth's interior)	Local effect	
Glacio-isostasy (loading or unloading of ice)	Local effect	10 mm/yr
Hydro-isostasy (loading or unloading of water)	Local effect	
Volcano-isostasy (magmatic extrusions)	Local effect	
Sediment-isostasy (deposition and erosion of sediments)	Local effect	< 4 mm/yr
Tectonic uplift/subsidence		
Vertical and horizontal motions of crust (in response to fault motions)	Local effect	1–3 mm/yr
Sediment compaction		
Sediment compression into denser matrix (particularly significant in and near river deltas)	Local effect	
Loss of interstitial fluids (withdrawal of groundwater or oil)	Local effect	≤ 55 mm/yr
Earthquake-induced vibration	Local effect	
Departure from geoid		
Shifts in hydrosphere, aesthenosphere, core-mantle interface	Local effect	
Shifts in earth's rotation, axis of spin and precession of equinox	Eustatic	
External gravitational changes	Eustatic	
Evaporation and precipitation (if due to a long-term pattern)	Local effect	

Changes Through Geologic Time

Sea level has changed over geologic time. As the graph shows, sea level today is very near the lowest level ever attained (the lowest level occurred at the Permian-Triassic boundary about 250 million years ago).

During the most recent ice age (at its maximum about 20,000 years ago) the world's sea level was about 130 m lower than today, due to the large amount of sea water that had evaporated and been deposited as snow and ice, mostly in the Laurentide ice sheet. Most of this had melted by about 10,000 years ago.

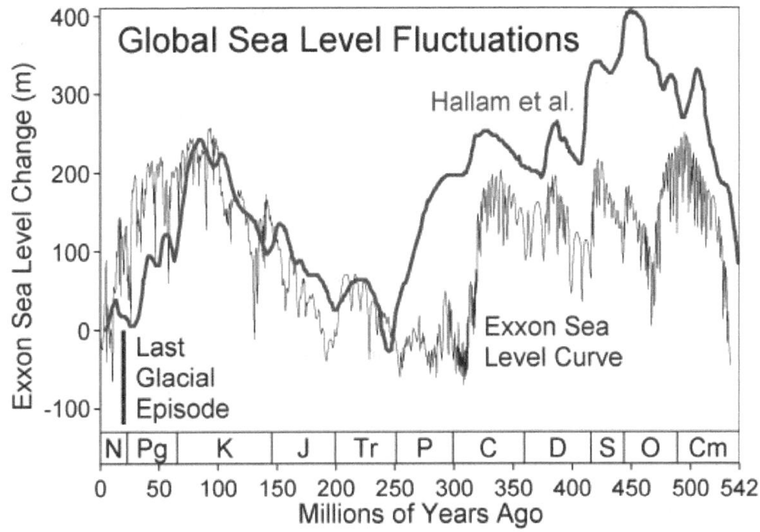

Figure: *Comparison of two sea level reconstructions during the last 500 Ma. The scale of change during the last glacial/interglacial transition is indicated with a black bar. Note that over most of geologic history long-term average sea level has been significantly higher than today.*

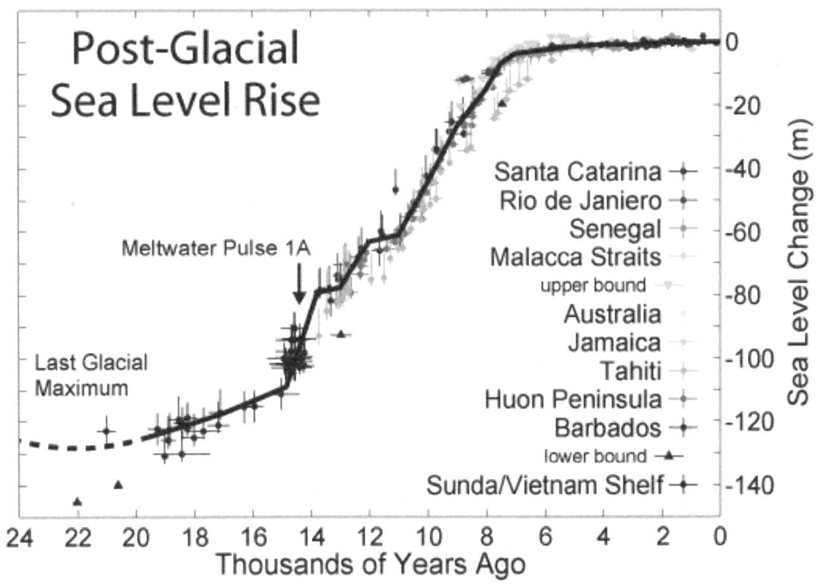

Figure: *Sea level change since the end of the last glacial episode. Changes displayed in metres.*

Hundreds of similar glacial cycles have occurred throughout the Earth's history. Geologists who study the positions of coastal sediment deposits through time have noted dozens of similar basinward shifts of shorelines associated with a later recovery. This results in sedimentary cycles which in some cases can be correlated around the world with great confidence. This relatively new branch of geological science linking eustatic sea level to sedimentary deposits is called sequence stratigraphy.

The most up-to-date chronology of sea level change during the Phanerozoic shows the following long term trends:

- Gradually rising sea level through the Cambrian
- Relatively stable sea level in the Ordovician, with a large drop associated with the end-Ordovician glaciation
- Relative stability at the lower level during the Silurian
- A gradual fall through the Devonian, continuing through the Mississippian to long-term low at the Mississippian/ Pennsylvanian boundary
- A gradual rise until the start of the Permian, followed by a gentle decrease lasting until the Mesozoic.

Recent Changes

For at least the last 100 years, sea level has been rising at an average rate of about 1.8 mm (0.1 in) per year. Most of this rise can be attributed to the increase in temperature of the sea and the resulting slight thermal expansion of the upper 500 metres (1,640 feet) of sea water. Additional contributions, as much as one-fourth of the total, come from water sources on land, such as melting snow and glaciers and extraction of groundwater for irrigation and other agricultural and human needs.

Current sea level rise is about 3 mm/year worldwide. According to the US National Oceanic and Atmospheric Administration (NOAA), "this is a significantly larger rate than the sea-level rise averaged over the last several thousand years", and the rate may be increasing. Sea level rises can considerably influence human populations in coastal and island regions and natural environments like marine ecosystems.

Between 1870 and 2004, global average sea levels rose a total of 195 mm (7.7 in), and 1.46 mm (0.057 in) per year. From 1950 to 2009, measurements show an average annual rise in sea level of 1.7 ± 0.3 mm per year, with satellite data showing a rise of 3.3 ± 0.4 mm per year from 1993 to 2009, The reason for recent increase is unclear, perhaps owing to decadal variation. It is unclear whether the increased rate reflects an increase in the underlying long-term trend.

There are two main mechanisms that contribute to observed sea level rise: (1) thermal expansion: ocean water expands as it warms; and (2) the melting of major stores of land ice like ice sheets and glaciers.

Sea level rise is one of several lines of evidence that support the view that the global climate has recently warmed. The global community of climate scientists confirms that it is very likely human-induced (anthropogenic) warming contributed to the sea level rise observed in the latter half of the 20th century.

Sea level rise is expected to continue for centuries. In 2007, the Intergovernmental Panel on Climate Change (IPCC) projected that during the 21st century, sea level will rise another 18 to 59 cm (7.1 to 23.2 in), but these numbers do not include "uncertainties in climate-carbon cycle feedbacks nor do they include the full effects of changes in ice sheet flow". More recent projections assessed by the US National Research Council (2010) suggest possible sea level rise over the 21st century of between 56 and 200 cm (22 and 79 in). The Third National Climate Assessment (NCA), released May 6th, 2014, projects a sea level rise of 1 to 4 feet by 2100(30-120 cm).

On the timescale of centuries to millennia, the melting of ice sheets could result in even higher sea level rise. Partial deglaciation of the Greenland ice sheet, and possibly the West Antarctic ice sheet, could contribute 4 to 6 m (13 to 20 ft) or more to sea level rise.

Work by a team led by Veerabhadran Ramanathan of the Scripps Institution of Oceanography suggests that a quick way to stave off impending sea level rise is to cut emissions of short-lived climate warmers such as methane and soot.

Aviation

Using pressure to measure altitude results in two other types of altitude. Distance above *true* or *MSL* (mean sea level) is the next best measurement to absolute. Above mean sea level is abbreviated as AMSL. MSL altitude is the distance above where sea level would be if there were no land. If one knows the elevation of terrain, the distance above the ground is calculated by a simple subtraction.

An MSL altitude—called pressure altitude by pilots—is useful for predicting physiological responses in unpressurised aircraft (see hypoxia). It also correlates with engine, propeller and wing performance, which all decrease in thinner air.

Pilots can estimate height above terrain with an altimeter set to a defined barometric pressure. Generally, the pressure used to set the altimeter is the barometric pressure that would exist at MSL in the region being flown over.

This pressure is referred to as either QNH or "altimeter" and is transmitted to the pilot by radio from air traffic control (ATC) or an Automatic Terminal Information Service (ATIS). Since the terrain elevation is also referenced to MSL, the pilot can estimate height above ground by subtracting the terrain altitude from the altimeter reading.

Aviation charts are divided into boxes and the maximum terrain altitude from MSL in each box is clearly indicated. Once above the transition altitude, the altimeter is set to the international standard atmosphere (ISA) pressure at MSL which is 1013.25 hPa or 29.92 inHg.

Flight Level

MSL is useful for aircraft to avoid terrain, but at high enough altitudes, there is no terrain to avoid. Above that level, pilots are primarily interested in avoiding each other, so they adjust their altimeter to standard temperature and pressure conditions (average sea level pressure and temperature) and disregard actual barometric pressure—until descending below transition level.

To distinguish from MSL, such altitudes are called flight levels. Standard terminology is to express flight level as hundreds of feet, so FL 240 is 24,000 feet (7,300 m). Pilots use the international standard pressure setting of 1013.25 hPa (29.92 inHg) when referring to flight levels. The altitude at which aircraft are mandated to set their altimeter to flight levels is called "transition altitude". It varies from country to country. For example in the U.S. it is 18,000 feet, in many European countries it is 3,000 or 5,000 feet.

Effects of Sea-level Rise

Based on the projected increases stated above, the IPCC TAR WGII report (*Impacts, Adaptation Vulnerability*) notes that current and future climate change would be expected to have a number of impacts, particularly on coastal systems.

Such impacts may include increased coastal erosion, higher storm-surge flooding, inhibition of primary production processes, more extensive coastal inundation, changes in surface water quality and groundwater characteristics, increased loss of property and coastal habitats, increased flood risk and potential loss of life, loss of non-monetary cultural resources and values, impacts on agriculture and aquaculture through decline in soil and water quality, and loss of tourism, recreation, and transportation functions.

There is an implication that many of these impacts will be detrimental—especially for the three-quarters of the world's poor who depend on agriculture systems. The report does, however, note that owing to the great diversity of coastal environments; regional and local differences in projected relative sea level and climate changes; and differences in the resilience and adaptive capacity of ecosystems, sectors, and countries, the impacts will be highly variable in time and space.

Statistical data on the human impact of sea-level rise is scarce. A study in the April, 2007 issue of *Environment and Urbanization* reports that 634 million people live in coastal areas within 30 feet (9.1 m) of sea level. The study also reported that about two thirds of the world's cities with over five million people are located in these low-lying coastal areas. The IPCC report of 2007 estimated that accelerated melting of the Himalayan ice caps and the resulting rise in sea levels would likely increase the severity of flooding in the short term during the rainy season and greatly magnify the impact of tidal storm surges during the cyclone season. A sea-level rise of just 400 mm in the Bay of Bengal would put 11 percent of the Bangladesh's coastal land underwater, creating 7–10 million climate refugees.

Island Nations

IPCC assessments suggest that deltas and small island states are particularly vulnerable to sea-level rise caused by both thermal expansion and ocean volume. Sea level changes have not yet been conclusively proven to have directly resulted in environmental, humanitarian, or economic losses to small island states, but the IPCC and other bodies have found this a serious risk scenario in coming decades.

Many media reports have focused on the island nations of the Pacific, notably the Polynesian islands of Tuvalu, which based on more severe flooding events in recent years, were thought to be "sinking" due to sea level rise. A scientific review in 2000 reported that based on University of Hawaii gauge data, Tuvalu had experienced a negligible increase in sea level of 0.07 mm a year over the past two decades, and that ENSO had been a larger factor in Tuvalu's higher tides in recent years. A subsequent study by John Hunter from the University of Tasmania, however, adjusted for ENSO effects and the movement of the gauge (which was thought to be sinking). Hunter concluded that Tuvalu had been experiencing sea-level rise of about 1.2 mm per year. The recent more frequent flooding in Tuvalu may also be due to an erosional loss of land during and following the actions of 1997 cyclones Gavin, Hina, and Keli.

Besides the issues that flooding brings (soil salinisation, ...) for these islands states, the islands states themselves would also become dissolved over time, as the islands becomes uninhabitable or becomes completely submerged by the sea. Once this happens, all rights on the surrounding area (sea) are removed. This area can be huge as rights extend to a radius of 224 nautical miles (414 km) around the entire island state. Any resources (fossil oil, minerals, metals, ...) within this area can be freely dug up by anyone and sold without needing to pay any commission to the (now dissolved) island state.

Numerous options have been proposed that would assist island nations to adapt to rising sea level.

Cities

Miami has been listed as "the number-one most vulnerable city worldwide" in terms of potential damage to property from storm-related flooding and sea-level rise.

Satellite Sea Level Measurement

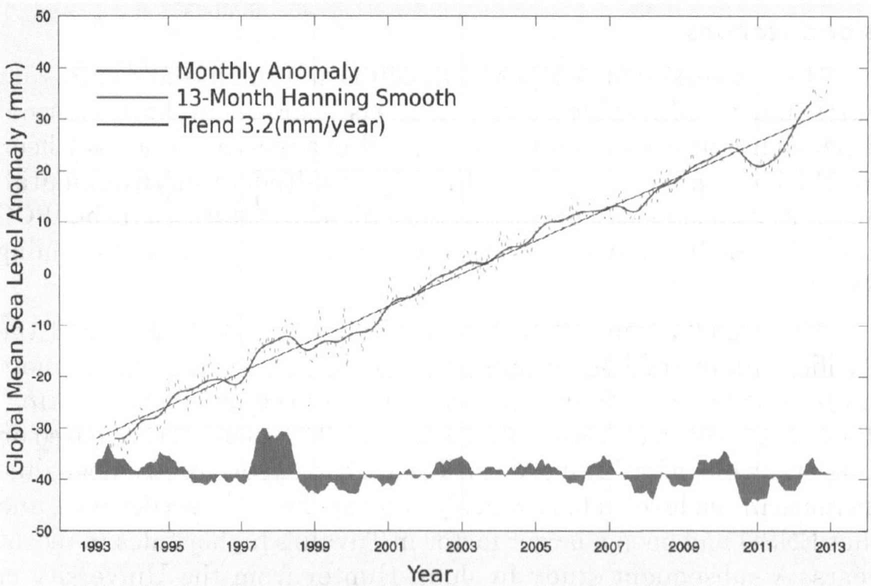

Figure: *Satellite Measurement of Sea Level.*

Current rates of sea level rise from satellite altimetry have been estimated in the range of 2.9–3.4 ± 0.4–0.6 mm per year for 1993–2010. This exceeds those from tide gauges. It is unclear whether this represents an increase over the last decades; variability; true differences between satellites and tide gauges; or problems with satellite calibration.

Knowing the current altitude of a satellite which can measure sea level to a precision of about 20 millimetres (e.g. the Topex/Poseidon system) is primarily complicated by orbital decay and the difference between the assumed orbit and the earth geoid. This problem is partially corrected by regular re-calibration of satellite altimeters from land stations whose height from MSL is known by surveying. Over water, the height is calibrated from tide gauge data which is needed to correct for tides and atmospheric effects on sea level.

Chapter 4

Subtidal and Beach Ecosystems

For MPA monitoring, soft-bottom subtidal ecosystems encompass the areas of sediment substrate occurring at depths between mean lower low water and 100 meters. These habitats are much more common than rocky habitats at all depth zones throughout the California coast. Although these habitats are seemingly simple and unstructured, they are rather dynamic, and inhabitant species must contend with dramatic changes as waves and currents shift sand and sediment across large areas.

A Home in Deep Groves

Adding to the complexity of these habitats, some areas feature rippled scour depressions (RSDs), or deposits of coarse-grained sediments that are depressed below the surrounding sediment that form important habitat for fish and invertebrates. Brittle stars and sea stars are common in this habitat, as well as commercially important species such as Dungeness crab, sand dabs, and starry flounder.

Understanding Trends Inside and Outside Mpas

Many of the fish and invertebrate species in soft-bottom subtidal ecosystems are wide-ranging, and individuals are likely to move between protected and unprotected locations. Detecting the effects of MPA designation on these species is challenging, but insights can be garnered by combining ecological data with information about the spatial patterns of fishing occurring outside of MPAs.

Decadal-scale shifts in the California Current can also result in warm and cool regimes that affect the sediment-inhabiting communities in this ecosystem. When warm regimes occur, declines in plankton production are likely – resulting in species and community declines as well. On shorter timescales, El Niño events, which increase wave activity and storms (leading to sedimentation), can cause major short-term disturbances to these ecosystems. As a result, integrated analyses of trend data are needed to separate MPA effects from other anthropogenic and natural system drivers.

Rocky Shores

A rocky shore is an intertidal area that consists of solid rocks. It is often a biologically rich environment and can include many different habitat types like steep rocky cliffs, platforms, rock pools and boulder fields. Because of the continuously action of the tides, it is characterized by erosional features.

Figure: *The rise and fall of tides on a rocky shore can define a volatile habitat for marine life*

Together with the wind, sunlight and other physical factors it creates a complex environment. Organisms that live in this area experience daily fluctuations in their environment. For this reason, they must be able to tolerate extreme changes in temperature, salinity, moisture and wave action to survive.

Formation

Sufficient loose or unconsolidated material and a suitable coastal environment to allow the sediments to accumulate are key features in the formation of beaches. The sediments can be terrigenous (land-derived), transported by rivers and streams or through erosion of coastal cliffs. In tropical regions or places with a lack of land-derived material, sediments may be biogenous and consists of broken corals or shells. Beaches on volcanic islands may be black and consists of broken lava and volcanic minerals.

Rocky shores are usually steeper than sandy shores. These differences are the result of the different permeability of the sediments and the balance between waves that retreat and come up. Rocky shores have a high permeability and much of the upcoming wave percolates into the bottom. This reduces the outwash of sediment toward the sea. For this reason, coarse sediment can accumulate on the beach and can be transported further on the beach by stronger waves.

***Figure:** Rocky shore in Lanzarote, Spain*

Zonation

Marine biologists divide the intertidal region into three zones (low, middle, and high), based on the overall average exposure of the zone. The low intertidal zone, which borders on the shallow subtidal zone, is only exposed to air at the lowest of low tides and is primarily marine in character. The mid intertidal zone is regularly exposed and submerged by average tides. The high intertidal zone is only covered by the highest of the high tides, and spends much of its time as terrestrial habitat. The high intertidal zone borders on the splash zone (the region above the highest still-tide level, but which receives wave splash). On shores exposed to heavy wave action, the intertidal zone will be influenced by waves, as the spray from breaking waves will extend the intertidal zone.

Depending on the substratum and topography of the shore, additional features may be noticed. On rocky shores, tide pools form in depressions that fill with water as the tide rises. Under certain conditions, such as those at Morecambe Bay, quicksand may form.

Figure: *Tide pools at Pillar Point showing zonation on the edge of the rock ledge*

Each region on the coast has a specific group of organisms that form distinct horizontal bands or zones on the rocks. The appearance of dominant species in these zones is called vertical zonation. It is a nearly universal feature of the intertidal zone.

Supratidal Zone

When the tide retreats, the upper regions become exposed to air. The organisms that live in this region are facing problems like gas exchange, desiccation, temperature changes and feeding. This upper region is called the supratidal or splash zone. It is only covered during storms and extremely high tides and is moistened by the spray of the breaking waves. Organisms are exposed to the drying heat of the sun in the summer and to extreme low temperatures in the winter. Because of these severe conditions, only a few resistant organisms live here.

Common organisms are lichens. They are composed of fungi and microscopic algae living together and sharing food and energy to grow. The fungi trap moisture for both themselves and their algal symbiont. The algae on the other hand produce nutrients by photosynthesis. Green algae and cyanobacteria can also be found on the rocks of the North Atlantic coasts. They are capable of surviving on the moisture of the sea spray from waves.

During the winter, they are found lower on the intertidal rocks. The algae growing higher on the rocks gradually die when the air temperature changes. At the lower edge of the splash zone, rough snails (periwinkles) graze on various types of algae. These snails are well adapted to life out of the water by trapping water in their mantle cavity or hiding in cracks of rocks. Other common animals are isopods, barnacles, limpets,

Subtidal Zone

The subtidal zone or sublittoral zone is the region below the intertidal zone and is continuously covered by water. This zone is much more stable than the intertidal zone. Temperature, water pressure and sunlight radiation remain nearly constant. Organisms do not dry out as often as organisms higher on the beach. They grow much faster and are better in competition for the same niche. More essential nutrients are acquired from the water and they are buffered from extreme changes in temperature.

Intertidal Zone

The intertidal zone, also known as the foreshore and seashore and sometimes referred to as the littoral zone, is the area that is above water at low tide and under water at high tide (in other words, the area between tide marks). This area can include many different types of habitats, with many types of animals, such as starfish, sea urchins, and numerous species of coral. The well-known area also includes steep rocky cliffs, sandy beaches, or wetlands (e.g., vast mudflats). The area can be a narrow strip, as in Pacific islands that have only a narrow tidal range, or can include many meters of shoreline where shallow beach slopes interact with high tidal excursion.

Organisms in the intertidal zone are adapted to an environment of harsh extremes. Water is available regularly with the tides but varies from fresh with rain to highly saline and dry salt with drying between tidal inundations. The action of waves can dislodge residents in the littoral zone. With the intertidal zone's high exposure to the sun, the temperature range can be anything from very hot with full sun to near freezing in colder climates. Some microclimates in the littoral zone are ameliorated by local features and larger plants such as mangroves. Adaptation in the littoral zone allows the use of nutrients supplied in high volume on a regular basis from the sea which is actively moved to the zone by tides. Edges of habitats, in this case land and sea, are themselves often significant ecologies, and the littoral zone is a prime example.

Figure: *Intertidal zonation: at low tide, the 3 typical intertidal zones can be seen*

A typical rocky shore can be divided into a spray zone or splash zone (also known as the supratidal zone), which is above the spring high-tide line and is covered by water only during storms, and an intertidal zone, which lies between the high and low tidal extremes. Along most shores, the intertidal zone can be clearly separated into the following subzones: high tide zone, middle tide zone, and low tide zone. The intertidal zone is one of a number of marine biomes or habitats, including estuaries, neritic, surface and deep zones.

The intertidal zone can be divided in three zones:

- High tide zone or high intertidal zone. This region is only flooded during high tides. Organisms that you can find here are anemones, barnacles, chitons, crabs, isopods, mussels, sea stars, snails,...

- Middle tide zone or mid-littoral zone. This is a turbulent zone that is (un)covered twice a day. The zone extends from the upper limit of the barnacles to the lower limit of large brown algae (e.g. *Laminariales, Fucoidales*). Common organisms are snails, sponges, sea stars, barnacles, mussels, sea palms, crabs,...
- Low intertidal zone or lower littoral zone. This region is usually covered with water. It is only uncovered when the tide is extremely low. In contrast to the other zones, the organisms are not well adapted to long periods of dryness or to extreme temperatures. The common organisms in this region are brown seaweed, crabs, hydroids, mussels, sea cucumber, sea lettuce, sea urchins, shrimps, snails, tube worms,...

Tidal pools are rocky pools in the intertidal zone that are filled with seawater. They are formed where hard or resistant rocks occur and where wave action and other erosional processes have eroded parts of the exposed rocks.

This leaves holes or depressions in the substrate where seawater can be collected at high tide. It is important that the rocks are hard enough to survive the regular wave actions, but be sufficiently weak (fractures, bedding differences) to be partly eroded. They can be small and shallow or deep.

The smallest ones are usually found at the high intertidal zone, whereas the bigger ones are found near the water. When the tide retreats, the pool becomes isolated. Because of the regular tides, the pool is not stagnant and new water regularly enters the pool. This is necessary to avoid temperature stress, salinity stress, nutrient stress,... Pools that are located higher on the beach are not regularly renewed by tides. These pools are basically freshwater or brackish water communities. It has different characteristics in comparison with other coastal habitats. Several taxa are more abundant in pools than the surrounding environment. These taxa are members of the algae and gastropods. There is also a difference between high and low located pools for the composition. In low located pools, whelks, mussels, sea urchins and *Littorina littorea* are common. Periwinkles and *Littorina rudis* are found in high located pools. Other organisms that are commonly found in pools are flatworms, rotifers, cladocerans, copepods, ostracods, barnacles, amphipods, isopods, chironomid larvae and oligochaetes. Vertical zonation also has been documented in tidal pools.

Littoral Zone

The littoral zone is the part of a sea, lake or river that is close to the shore. In coastal environments the littoral zone extends from the high water mark, which is rarely inundated, to shoreline areas that are permanently submerged. It always includes this intertidal zone and is often used to mean the same as the intertidal zone. However, the meaning of "littoral zone" can extend well beyond the intertidal zone.

There is no single definition. What is regarded as the full extent of the littoral zone, and the way the littoral zone is divided into subregions, varies in different contexts (lakes and rivers have their own definitions). The use of the term also varies from one part of the world to another, and between different disciplines. For example, military commanders speak of the littoral in ways that are quite different from marine biologists.

The adjacency of water gives a number of distinctive characteristics to littoral regions. The erosive power of water results in particular types of landforms, such as sand dunes, and estuaries. The natural movement of the littoral along the coast is called the littoral drift. Biologically, the ready availability of water enables a greater variety of plant and animal life, and particularly the formation of extensive wetlands. In addition, the additional local humidity due to evaporation usually creates a microclimate supporting unique types of organisms.

The word "littoral" is used both as a noun and an adjective. It derives from the Latin noun *litus, litoris*, meaning "shore". (The doubled 't' is a late medieval innovation and the word is sometimes seen in the more classical-looking spelling 'litoral'.)

In oceanography and marine biology, the idea of the littoral zone is extended roughly to the edge of the continental shelf. Starting from the shoreline, the littoral zone begins at the spray region just above the high tide mark. From here, it moves to the intertidal region between the high and low water marks, and then out as far as the edge of the continental shelf. These three subregions are called, in order, the supralittoral zone, the eulittoral zone and the sublittoral zone.

Eulittoral Zone

The *eulittoral zone* (also called the *midlittoral* or *mediolittoral zone*) is the *intertidal zone* also known as the *foreshore*. It extends from the spring high tide line, which is rarely inundated, to the spring low tide line, which is rarely not inundated. The wave action and turbulence of recurring tides shapes and reforms cliffs, gaps, and caves, offering a huge range of habitats for sedentary organisms. Protected rocky shorelines usually show a narrow almost homogenous eulittoral strip, often marked by the presence of barnacles. Exposed sites show a wider extension and are often divided into further zones.

Sublittoral Zone

The *sublittoral zone* starts immediately below the eulittoral zone. This zone is permanently covered with seawater and is approximately equivalent to the neritic zone.

In physical oceanography, the sublittoral zone refers to coastal regions with significant tidal flows and energy dissipation, including non-linear flows, internal waves, river outflows and oceanic fronts. In practice, this typically extends to the edge of the continental shelf, with depths around 200 meters.

In marine biology, the sublittoral refers to the areas where sunlight reaches the ocean floor, that is, where the water is never so deep as to take it out of the photic zone. This results in high primary production and makes the sublittoral zone the location of the majority of sea life. As in physical oceanography, this zone typically extends to the edge of the continental shelf. The benthic zone in the sublittoral is much more stable than in the intertidal zone; temperature, water pressure, and the amount of sunlight remain fairly constant. Sublittoral corals do not have to deal with as much change as intertidal corals. Corals can live in both zones, but they are more common in the sublittoral zone.

Within the sublittoral, marine biologists also identify the following:

- The infralittoral zone is the algal dominated zone to maybe five metres below the low water mark.
- The circalittoral zone is the region beyond the infralittoral, that is, below the algal zone and dominated by sessile animals such as oysters.

Shallower regions of the sublittoral zone, extending not far from the shore, are sometimes referred to as the subtidal zone.

Supralittoral Zone

The *supralittoral zone* (also called the *splash*, *spray* or *supratidal zone*) is the area above the spring high tide line that is regularly splashed, but not submerged by ocean water. Seawater penetrates these elevated areas only during storms with high tides. Organisms here must cope also with exposure to fresh water from rain, cold, heat and predation by land animals and seabirds. At the top of this area, patches of dark lichens can appear as crusts on rocks. Some types of periwinkles, Neritidae and detritus feeding Isopoda commonly inhabit the lower supralittoral.

In Freshwater Ecosystems

In freshwater situations, littoral zones occur on the edge of large lakes and rivers, often with extensive areas of wetland. Hence, they are sometimes referred to as fringing wetlands. Here, the effects of tides are minimal, so other definitions of "littoral" are used. For example, the Minnesota Department of Natural Resources defines littoral as that portion of the lake that is less than 15 feet in depth.

The littoral zone may form a narrow or broad fringing wetland, with extensive areas of aquatic plants sorted by their tolerance to different water depths. Typically, four zones are recognized, from higher to lower on the shore: wooded wetland, wet meadow, marsh and aquatic vegetation. The relative areas of these four types depends not only on the profile of the shoreline, but upon past water levels. The area of wet meadow is particularly dependent upon past water levels; in general, the area of wet meadows along lakes and rivers increases with natural water level fluctuations. Many of the animals in lakes and rivers are dependent upon the wetlands of littoral zones, since the rooted plants provide habitat and food. Hence, a large and productive littoral zone is considered an important characteristic of a healthy lake or river.

Littoral zones are at particular risk for two reasons. First, human settlement is often attracted to shorelines, and settlement often disrupts breeding habitats for littoral zone species. For example, many turtles are killed on roads when they leave the water to lay their eggs in upland sites. Fish can be negatively affected by docks and retaining walls which remove breeding habitat in shallow water. Some shoreline communities even deliberately try to remove wetlands since they may interfere with activities like swimming. Overall, the presence of human settlement has a demonstrated negative impact upon adjoining wetlands. An equally serious problem is the tendency to stabilize lake or river levels with dams. Dams removed the spring flood which carries nutrients into littoral zones, and reduces the natural fluctuation of water levels upon which many wetland plants and animals depend. Hence, over time, dams can reduce the area of wetland from a broad littoral zone to a narrow band of vegetation. Marshes and wet meadows are at particular risk.

Problems and Adaptations

The continuously changing environment makes that organisms have to be tolerant for these changes. Adaptations are a solution for these problems and are necessary to survive.

Air

Intertidal organisms are regularly exposed to air and water. Air differs physically from seawater in diverse and important features. This influences the ability to exchange gas and their overall thermal balance with the surrounding environment. The chemical and gaseous constituents of the atmosphere have changed dramatically through geological times. Because of the much greater density of water, organisms become buoyant. In air, the gravity induces retraction of tentacles and other feeding organs. It also makes the body less resistant. For this reason, organisms need supporting structures when they are exposed to air. Attachment and body changes are also required. When exposed to the air, the organisms directly absorb solar radiation. The buffering capacity of water, because of the high rate of heat conductivity, disappears and the body temperature increases. In contrast to this, heat loss is much lower in air than in water. An adaptation to heating is the vaporization of internal water reserves.

Light

Sunlight is another parameter that influences the organisms. When there is too much sunlight, organisms dry out and the capacity to capture light energy can be weakened. The light that is not used or dissipated can cause damage to subcellular structures. Too little sunlight reduces the growth and reproduction of the organism, because photosynthesis is reduced. Algae can avoid absorbing too much light by changing the complement or amount of pigments they produce. They also can rearrange the pigmented organelles within their cells. When free radicals are produced from an excess of light, they can be scavenged and deactivated.

Temperature

The intertidal zone can experience extreme temperature changes. The organisms in this zone must be resistant to these changes to survive. Most of the marine organisms are ectothermic and need the warmth from the environment to survive. When the organisms are submerged, they are buffered against temperature changes, because the water is isothermal. When the organisms become exposed to the air, they can experience cool or warm temperatures. When the temperature is too low, the organisms must cope with physiological threats associated with cold stress. This can be the case in polar and temperate latitude coastal zones. The body fluids can then reach their freezing point and ice crystals develop. This causes damage to cell membranes and increasing the osmotic concentration of the remaining fluids. To avoid this cold stress, organisms can migrate to habitats that are more suitable. This can be a problem for sessile organisms. They can develop physiological and behavioral adaptations such as gaping shells (mussels).

Some organisms have developed antifreeze proteins. Increasing the concentration of small osmolytes such as glycerol in the body fluids can decrease the freezing point. Another strategy is to control ice crystal formation. Organisms can control the speed and the exact location of the ice crystals. When the ice formation is intracellular, it is lethal but extracellular ice formation can be tolerated. When the temperature is too high, heat stress appears. Heat stress accelerates rates of metabolic processes. This can be avoided by evaporative cooling combined with circulation of body fluids.

Salinity Stress

Salinity stress can occur in the external medium and in surface films. The concentration of the fluids determines whether or not the organism will lose water. When the osmolality of the cell is lower than the surrounding medium, the cell loses water from the internal fluids to the environment (hyperosmotic stress). When the intracellular osmolality is higher than the environment, there is an influx of water into the cell from the environment (hypoosmotic stress). Multicellular organisms respond to this salinity stress by compartmentalization. This buffers the cells from sharp changes in the osmotic environment. When the tissue has an immediate contact with the external medium, a solution can be to regulate intercellular osmotic pressure by actively excreting salts or water. Another solution is to change the internal osmolality. This can be done by incorporating ions or compatible solutes in the internal fluids.

Desiccation Stress

Organisms are threat by desiccation during emersion at low tides or when they are positioned in the high intertidal zones. Dehydration due to evaporative water loss is the most common mechanism. Highly mobile organisms can avoid the desiccation by migrating to a region that is more suitable. Less mobile organisms restrict various activities (reduced metabolism) and attach more firmly to the substrate. Physiological features to tolerate water loss are desiccation-resistant egg cases, reduction in water permeability of membranes, accumulation of metabolic end products, reduction of metabolic and developmental rates, maintenance of intracellular osmolytes and gene expression for production of protective macromolecules.

Predation

A wide variety of strategies to escape from predation exists. The first strategy is calcification. It makes it more difficult for the predator to eat these organisms. This strategy is applied by algae. It makes them tougher and less nutritious. A second one is the production of chemicals, usually produced as secondary metabolites. These chemicals can be produced all the time such as toxins, but other chemicals are only produced in response to stimuli (inducible defence). Another way to avoid predation is to have two distinct anatomical forms within one life cycle. This can be e.g. an alternation between a crusty form when the predator is present and a more delicate form (e.g. blade) when the predator is absent. Also the shape of the body can be a distinct evolutionary advantage. Bioluminescence is another strategy to avoid predators. Many intertidal and subtidal predators visually forage. The light is used for warning, blinding, making scare, misleading or attracting the predator. A commonly used form of protection against predation is camouflage. This can be visually or chemically. Visual camouflage means that the prey becomes invisible to the predator by using the same colors as the environment. Chemical camouflage is the passive adsorption of chemicals. The predator does not smell the prey anymore, because the smell is masked.

Wave Action

One way to protect organisms from waves is permanent attachment. But this strategy can not be used by organisms that have to move to feed themselves. These organisms have to make a compromise between mobility and attachment. Another way to be protected is to burrow themselves into the sediment. But an alternative is to seek protected habitats. Attachment can be done by different structures. Bivalves usually use threads (byssal threads) to attach to rocky surfaces or to other organisms. But it can also be done by a foot. Another one is cementation. This is the case for bivalves such as oysters, scallops and some other forms. They lay on their side, with the lower valve cemented firmly to the bottom. This can be combined by reduction or enlargement of certain muscles.

Flora and Fauna

Aside from the striking natural beauty that plants provide, they are crucial parts of beach ecosystems. Plants are among the first forms of life to inhabit beaches and dunes. Once plants have taken root, sands begin to stabilize, slowly at first, then faster as more plants take root. Plants serve as habitat for many animal species from providing basic shade to providing nest material or even allowing a home to be built inside them. They offer camouflage to animals and are a primary food source for some. Without plants, beaches would probably be devoid of animal life.

Figure: *Large, beautiful, vegetated dune in northern Morocco. Plants serve as a major part of the ecosystem here providing food for birds and small mammals.*

Beach Vines

Railroad vine, or bay hops (*Ipomoea pes-caprae*), is an important vine that stabilizes beach dunes with runners that can reach lengths of over 30 meters. This specialized morning glory can survive in drought, extreme heat, and very salty areas without problems. Without this plant, dunes would keep shifting location on the beach making the areas inhospitable to animals. But once this plant anchors the sand and begins to grow, other species such as sea oats, birds, and crabs are much more likely to thrive. Birds can make their nests between the runners and be camouflaged from predators. Other plants that require more stable soils can also move into areas where railroad vine lives stabilizing the dune even more. Stable dunes serve not only as wildlife habitat, but protect buildings and roads that may have been built too close to the ocean. This particular beach vine lives in the United States, but similar species perform similar ecosystem functions all over the world.

Beach Grass

Sea oats (*Uniola paniculata*) is an important plant species that stabilizes shifting sand dunes. Blowing sand is trapped by numerous stems of the plant. When sea oats are buried by shifting sand, the plants stops growing upward and instead grows horizontally just under the surface of the sand. This sometimes results in deeply buried underground stems that also help stabilize dunes. Because of this plant's importance, it is protected in many areas; in Georgia and Florida (US), it is illegal to pick sea oats.

Birds and the Beach

Beaches support an amazing variety of bird species from small sand pipers to large pelicans. Many birds spend large portions of their lives over the water, but they all must eventually come out of the air to find a mate and reproduce. The sand, rocks, pebbles and plants of the world's beaches serve as breeding grounds for many species of bird. Preservation of beaches as nesting sites is crucial for the protection of bird species. When beaches are developed, polluted, or destroyed, habitat critical to the life cycle of all of these animals is also destroyed. Beach sand mining is a major mechanism for habitat destruction. Below is some information on just a few species of birds that call the beach home.

Seagulls

Gulls are probably the animal people associate most with the beach, but many people know little about seagulls. These birds are extremely maneuverable able to hover in the slightest of winds and dive with amazing speed. There are over 50 species of seagulls ranging in size from about 30 cm to 75 cm with wingspans over 1.7 meters. Gulls are aggressive and will display such aggression when they are threatened. They scavenge for their food of crabs and fish and will steal food from their own species or others whenever possible. Gulls nest on the ground making habitat loss a major threat to their existence. Beach sand mining and development are among the worst threats to gull habitat. Although most gulls species are found in abundance, some species such as Olrog's Gull and Saunders' Gull are threatened with habitat loss as the single greatest threat. Other species are well adapted to life away from the coast with a few species prospering in garbage dumps. The abundance of these species camouflages the low numbers of other species perpetuating the myth that gulls will never run out of habitat. This is wrong. Beaches need to be protected to ensure the future of all gull species.

Piping Plover

The piping plover (*Charadrius melodus*) is native to the beaches of the North American Atlantic coast and the West Indies in the Gulf of Mexico. The plover is threatened in the United States and endangered in Canada. As a result, some efforts have been made to protect its habitat. This protection, however, would not be necessary if beaches were undeveloped or significantly less developed. In some areas, protection measures have been of limited benefit to the bird, while measures in other areas have been successful. Examples of protection measures range from protecting individual nests (located on the ground) to closing entire sections of beach.

The plover feeds on insects, worms, and crustaceans abundant in beach debris or wrack lines. It is a medium-sized bird with adult body lengths of around 17 cm and wingspans of up to 45 cm. It winters in the southern part of its range and returns to northern climes for breeding in the spring.

Pelicans

Pelicans live on every continent except Antarctica even though there are only about six species. This means that the bird is very well adapted to different conditions. One adaptation that pelicans share is the large pouch or gullet used to capture prey, usually fish. Another adaptation is the birds ability to live in a variety of conditions. For example, the American White Pelican (*Pelecanus erythrorhynchos*) lives inland much of the year and winters on the shores of the Gulf of Mexico. These large birds with wingspans of over 3 meters must eat over 1.5

kg of food per day. The Great White Pelican (*Pelecanus onocrotalus*) lives in Africa, Europe, and Asia. The bird frequents beaches in search of fish, crustaceans, and other small organisms on which to feed but lives most of its life on inland bodies of water.

Brown pelicans (*Pelecanus occidentalis*) are different from White Pelicans because Brown Pelicans live only on the beaches of North and South America making beach habitat crucial for the bird. Brown Pelicans were on the endangered species list in the United States for 40 years and only recently have been candidates for removal form the list. Brown Pelicans are much smaller than white ones but are still large birds. Wingspans are around 2 to 2.5 meters for brown pelicans.

Penguins

There are about 20 species of penguins, but only four species, the Adelie, the Emperor, the Chinstrap and the Gentoo, live on Antarctica. There are no penguins at the North Pole. The remaining species of penguins live in the southern hemisphere in Australia, New Zealand, Africa, and S. America with the exception of the Galapagos penguin, which lives near the equator. The Galapagos penguin thrives so far north because of the deep, cold-water Humboldt ocean current. All of these flightless birds spend their time in the cold ocean water and on the beach. Beaches are necessary for the penguins to make nests and lay their eggs. Without clean beaches, penguins all over the world would have difficulty surviving.

These curious, flightless birds are somewhat awkward on land, but in the water are extremely agile and adept swimmers. Larger species can dive to depths of over 500m and hold their breath for 15 minutes! Most penguins lay two eggs, but species in colder climates usually lay only one egg. The egg is small compared to the size of the mother. Penguins range in size from the 40 cm Little Blue penguin to the Emperor which can reach 1 m.

Terns

The birds that resemble seagulls but have orange beak are called terns and are related to sea gulls. Many species of terns have grey or black coloration on their heads too. There are over 40 species of terns and they live on every continent of the world. Terns generally eat small fish (which they spot while hovering over the water), bugs, or other small invertebrates. They range in size from the 22 cm long Least Tern (Sternula antillarum) to the 48 cm long Caspian Tern (*Hydroprogne caspia*).

Caspian terns (*Hydroprogne caspia*) are the largest terns in the world with wingspans of up to 140 cm and the most widely distributed tern; they are found on every continent except Antarctica. Like other terns, they hover over the water in search of fish. When they see one, they quickly dive by tucking in their wings. The beach is important to terns for nest sites. Without the beach, the terns wouldn't be able to find suitable sand, gravel, and vegetation to lay their eggs.

The Arctic tern (*Sterna paradisaea*) goes to the ends of the earth every year. It spends the summer of the southern hemisphere on Antarctica and then flies back to the Arctic Circle to breed during the northern hemisphere summer. This flight of over 18,000 km each year makes it the largest migration of any animal on earth! The Arctic Tern needs beaches for nesting sites. The birds dig shallow pits called scrapes in the sand or gravel laying 1-3 eggs. When they hatch, the young birds must learn to fly quickly for their long journey.

Other Beach Animals

Crabs: There are over 6,500 known crab species making them an extremely diverse group of animals. Crabs are covered in a hard exoskeleton which offers them protection from predators. As a group, they are scavengers with some species eating just about anything: algae, mollusks, worms, crustaceans, fungi, and other organic material. Crabs are an economically and culturally important species in many countries such as Japan and China. Crab habitat varies; some species live exclusively in the ocean in a range of water depths with a preference for shallow waters while some species live entirely on land. Some land species have structures that function as primitive lungs. Habitat loss is a major threat to many crab species as well as invasive and non-native species. Beach development, nourishment, and dredging produce fine sediment which is harmful to crabs as well.

Are Hermit Crabs True Crabs?: Hermit crabs are not very closely related to true crabs and live in shallow salt water in most areas of the world. This includes most beaches of the world meaning that beach destruction also threatens these tiny creatures too. There are about 500 species of hermit crabs. Hermit crabs can not make their own shells and crawl into shells from other sea creatures, most often sea snails. In order for the hermit crab to increase in size, a vacant shell must be available. If no shell is available, the hermit crab will not grow very quickly.

Crabs that we eat have an exoskeleton that they make themselves and are a much more diverse group than hermit crabs. There are over

6,500 species of true crabs and range in size from the Pea crab to the Japanese spider crab with a leg span of over 4 m!

Sea Turtles: Sea turtles are found in every ocean of the world except the Arctic Ocean even though there are only 7 species. All 7 species are threatened and 5 of them are endangered or critically endangered. These amazing swimmers are very similar to their ancestors and haven't changed much in over 10 million years. Among their adaptations are their highly evolved front flippers, ability to hold their breath for long periods of time, and hard shell. They are the only reptiles with a hard shell and some can live up to 80 years and weigh over 500 kg!

Figure: *Sea turtles in Hawaii. Photo: Mark Howdeshell/Jo Handinero*

There are numerous human threats to sea turtles. The shrimp industry represents the largest threat because turtles that end up caught in the long fishing nets usually die (see photo). Another threat is illegal poaching and harvesting of eggs and turtles for food. Turtle skins and shells are also used in arts and crafts. In some cultures, turtle eggs are believed to be an aphrodisiac. When traveling, do not support these industries and do not purchase souvenirs or goods made from these animals. Beach sand mining also represents a major threat to sea turtles. Beach sand mining in St. Lucia was extensive for the concrete industry until a moratorium was issued in 1984. A decade later in 1993, the beaches still had not recovered as turtles had to abandon nesting attempts when they hit the water table.

Sea Turtles and Beaches: Sea turtles use beaches as nesting sites for incubation of their eggs. Without nesting sites, it would be impossible for sea turtles to complete their life cycle. They spend much of their time eating sea grass and when the females are able to reproduce, they return to the beach to lay their eggs. In most cases, the females return to the beach where they were born using the earth's magnetic field to navigate. Once on the beach, they lay large clutches of eggs in shallow pits of sand where the young turtles mature. It takes around two months for the young turtles to mature. When the turtles finally hatch, they must crawl to the ocean quickly to avoid predation by birds, other animals, and human threats.

Loggerhead Sea Turtles: Loggerhead turtles (*Caretta caretta*) live in the Atlantic, Indian, and Pacific Oceans and reach up to 350 kg. Loggerheads have reddish brown shells and lighter skin. They are an endangered species meaning that the turtles are at risk of becoming extinct. Fishing nets, dredging sand off the ocean floor, and coastal development all threaten turtles and their habitat. If you spot a sea turtle in North America, it is probably a loggerhead because they are the most common turtle there.

Loggerhead turtles (*Caretta caretta*) live in the Atlantic, Indian, and Pacific Oceans and Like other sea turtles, the beach is necessary for loggerheads to survive. Loggerheads make their nests on the beaches of the Atlantic Ocean in the United States and in the Mediterranean Sea, usually laying between 75 and 150 eggs. If they survive the first few years of their lives, some may live to be over 75 years old!

Seals: These playful swimmers live all over the world and are intimate with our world's beaches. They are extremely well suited for life in the water with adaptations such as powerful tail muscles, warm blubber, and the ability to conserve oxygen on long dives. They feed in the water, but need beaches to breed, raise young, and reproduce. One the beach, they are also less exposed to predators like sharks.

What Exactly are Seals?: Seals are part of a group of animals called pinnipeds meaning "winged feet," which refers to their flippers. There are over 33 species of pinnipeds with three main divisions: true seals, seals with ears, and walruses. The true seals are more adapted to live in the water and can't "walk" on their hind flippers like the eared seals can. Pinnipeds are excellent swimmers and can hold their breath for almost 2 hours! They can do this by slowing their heart rate when they dive and changing the chemistry.

The beach is important to all pinnipeds for breeding and raising their young. Seals leave their young alone on the beach during the day while feeding. The mothers return at night to nurse their young. People often mistake baby seals as "orphans" on the beach that have been left by their parents, but this is not the case. Baby seals are sometimes left for days at a time while the mother is feeding. The best thing to do is to call local wildlife officials and remember that handling baby seals is illegal.

Microfossils

Microfossils are fossil that usually are smaller than four millimeters. From the White cliffs at Dover to the chalk caves of the Champagne region of France, these tiny fossils are found all over the world both on the coast and inland. Huge populations of tiny organisms with hard shells made of silica or calcium live in the world's oceans.

When these organisms die, their shells can become fossils and remain for very long periods of time. Although these organisms are very small, the fossilized remains of them can accumulate in huge quantities for two reasons:

1. there are so many of them, and
2. these groups of organisms have accumulated for millions of years.

When ocean temperatures are favorable for their life-cycles, deposits of their bodies on the ocean floor can reach amazing depths – hundreds of meters! These fossilized remains are referred to as oozes and there are two main types of oozes: calcareous (calcium-based) oozes and siliceous (silica-based) oozes.

Calcareous Oozes: Calcareous oozes, or calcium-based oozes, are made up of the shells of foraminifera (forams) and coccolithophores.

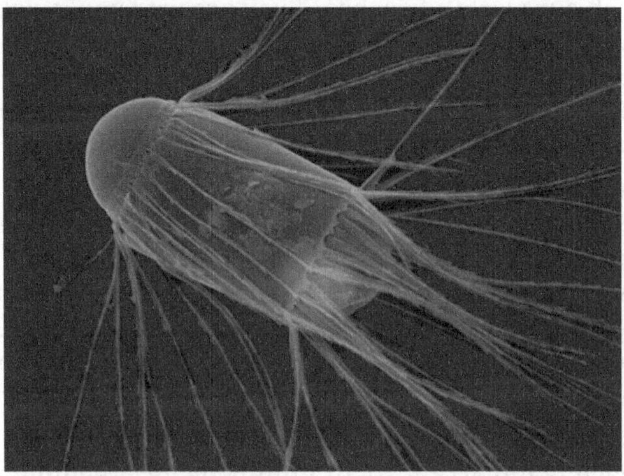

Figure: *Forams have shells made up of mostly calcium carbonate. Photo: USGS*

Forams are single-celled organisms that live only for about one month. These protists are very small, but they are very large in number: tens of thousands of species of forams live in the waters of the oceans. When forams die, their shells settle on the ocean floor and accumulate. Forams are very useful for dating rock layers since scientists have good data about which species of forams lived during particular time periods.

Coccolithophores are single celled organisms with shells made up mostly of calcium carbonate plates. When cocoliths die, their shells accumulate on the ocean floor. During the Cretaceous period some 145 million years ago, the oceans were warmer and cocoliths accumulated for very, very long periods of time. You probably are quite familiar with cocolithic rock, but you might not know it; chalk is an example.

Siliceous Oozes: Diatoms are a type of algae, most of which have one cell. There are a few hundred species alive today, but there are over 100,000 fossil species! These tiny algae use sunlight to make energy and account for nearly half of all the energy producing activities in the world's oceans (primary productivity – mostly through photosynthesis). Like the other oozes, when diatoms die, their small silica shells sink to the bottom of the ocean and slowly accumulate.

Radiolarians are a group of amoeba with a hard shell made of mostly silica. Like the other oozes listed above, the shells of these organisms accumulate on the ocean floor. You might know the word diatom if you have ever used soil with diatoms mixed in called diatomaceous earth. Other uses include pool filters, to clean up spills, and as an abrasive. It's a good things that we use diatoms for so many things – their silica shells never biodegrade!

Insects and Worms: In addition to larger plants and animals, beaches are also full of much smaller forms of life. Insects are present at virtually all beaches and provide important ecosystem functions such as breaking down organic matter, nutrient cycling, and serving as food sources for larger organisms. Many people only consider insects when their presence is in some way objectionable to their visit to the beach, such as mosquitoes, but these tiny creatures have some amazing adaptations that suit them to life on the beach. To find some insects on your beach, gently roll over some of the organic matter from the wrack line and look closely. Little is known about many of these species; who knows what you might find?

Northeastern beach tiger beetle

The Northeastern beach tiger beetle (*Cicindela dorsalis dorsalis*) is a beetle that, true to its name, is found on beaches in the northeastern

United States. Historically, it was found in every coastal state from Massachusetts to Virginia. Today, it is extirpated (or locally extinct) from Rhode Island, Connecticut, and New Jersey, with a small population remaining in Massachusetts and a larger population in the Chesapeake Bay. It is listed as endangered in Massachusetts, and threatened federally.

The ideal habitats for northeastern beach tiger beetles are long, wide, undisturbed, dynamic beaches. In the summer months, larvae live in the upper intertidal zone, and dig burrows up to 14 inches deep. Adults live between the dunes and the high-tide line. From about November to March of each year, larval northeastern beach tiger beetles overwinter up high on the beach, avoiding most waves and storm activity. After their hibernation, they will emerge and move back down to the intertidal zone.

Northeastern beach tiger beetles are handsome, sand-colored insects. Specifically, they are a whitish-tan color with dark markings and a green-bronze head. They are between ½ and ¾ inch long, with long slender legs and large pinching jaws. To catch food, larvae wait near the top of the burrow for small invertebrate prey to get close. Then, as "tiger-like" skilled predators, they grasp their prey (often lice, fleas, and flies) in a quick and aggressive manner, hence the name "tiger beetle." Once they emerge as adults, they can fly for up to a mile to find a more hospitable habitat or to scavenge for food such as dead fish and crabs for feeding.

Presence of the Northeastern beach tiger beetle is an indicator of a healthy beach.

Beach Hoppers: Beach hoppers, or sand fleas (*Orchestoidea* spp.), are common to the west coast of North America and are one of the most abundant beach insects. Relatives of the beach hopper can be found around the world. They spend most of the day buried head-first in the sand trying not to get carried away by the tides and most of the night feeding on kelp, other organic debris, and microorganisms living in the sand. If high tide does come at night when they are feeding, they take a break for several hours and burrow into the sand so they don't get swept out to sea. Move a pile of kelp on the beach, and you are likely to send hundreds of these tiny creatures hopping. Beach hoppers need sandy beaches for habitat, but are one of the few insects from the Amphipod order that require sandy beaches: nearly all the other Amphipods live in the ocean.

Kelp Flies: Several species of kelp fly live on the beaches on the west coast of North America and many more live around the world. These insects are responsible for cycling nutrients through the beach ecosystem and serve as a source of food for a variety of other animals. Kelp and seaweed wash up on most beaches and this decomposing organic matter serves as an ideal habitat for kelp flies.

Rove Beetles: Beetles are everywhere, including at the beach. The Beetle family (*Coleoptera*) is the largest family of animals on earth; over 350,000 species have been described with an estimated millions more undescribed. The rove beetles are a large portion of the beetle family with members living in nearly every conceivable habitat except the arctic regions. Rove beetles that live in beach sand have the ability to survive submerged for periods of time enabling some species to survive a high tide and then come out to feed when the tide recedes. Rove beetles feed on kelp fly larvae and pupae and other amphipods.

Worms: Beach sand seems like an unlikely place to find worms. The soil has little organic matter compared to soils away from the coast, the salt content can be very high, and moisture levels change dramatically as the tide changes. Despite all of these conditions some worms have adapted through time to thrive here.

Blood Worms: Blood worms from the genus Euzonus live all over the world, often in huge numbers. The fossil record shows a wide variety of species throughout history and a new species was observed in Japan in 2003. These specialized worms usually live in the top 20 cm of sand where oxygen supplies are plentiful in the intertidal zone. These amazing worms eat sand to gain nutrients from bacteria and algae living on the grains of sand. Imagine being able to survive from eating sand! They get their red color from red blood cells, the same kind that humans have, which enable them to live in low oxygen environments.

Sandy Shores

Sandy shores or beaches are loose deposits of sand, gravel or shells that cover the shoreline in many places. They make up two-thirds of the world's ice-free coastlines. Beaches serve as buffer zones or shock absorbers that protect the coastline, sea cliffs or dunes from direct wave attack. It is an extremely dynamic environment where sand, water and air are always in motion. Beaches also provide important coastal recreational areas for a many people. Fine-grained sand beaches tend to be quite flat.

Formation

Sandy beaches are soft shores that are formed by deposition of particles that have been carried by water currents from other areas. The transported material is in part derived from the erosion of shores, but the major part is derived from the land and transported by rivers to the sea. The two main types of beach material are quartz (=silica) sands of terrestrial origin and carbonate sands of marine origin. The carbonate sand is weathered from mollusk shells and skeletons of other animals. Other material includes heavy minerals, basalt (=volcanic origin) and feldspar.

Characteristics

The grain size of sand varies from very fine to very coarse. The particle diameter is shown in the table below. As said before, the two main types of beach material are quartz (=silica) sands of terrestrial origin and carbonate sands of marine origin. Quartz sands have a slightly lower density ($2.66 g.cm^{-3}$) than carbonate sands ($2.7 to 2.95 g.cm^{-3}$). The quartz particles also seem to be more rounded. Calcium carbonate particles sink more slowly in water due to their more irregular shapes, even if their density is higher.

Generic Name	*Particle Diameter (mm)*
Very coarse	1.0 to 2.0
Coarse	0.50 to 2.0
Medium	0.25 to 0.50
Fine	0.125 to 0.50
Very Fine	0.0625 to 0.125

Porosity is the volume of void space in the sand. It is the volume of water needed to saturate a given weight of dry sand. Most sands have a porosity of about 30 to 40 % of the total volume. The finer a sand the greater its porosity. Permeability is the rate of flow or drainage of water through the sand. Fine sands have lower permeabilities due to their smaller pore sizes. Penetrability is related to particle size and porosity. It can be important to the macrofauna. All species must be able to burrow into the substratum. To determine the penetrability, the proportion of clay and silt and the water content are very important.

The two basic beach types are dissipative and reflective. Together with the intermediate types, there are six major microtidal beach types. The reflective type occurs when conditions are calm and/or the sediment is coarse. There is no surf zone and waves flow upon the beach. It reflects a major part of the incoming wave. When bigger waves cut back a beach and spread out its sediments to form a surf zone, the reflective beaches create a series of intermediate types. When wave action is strong and/or sediment particle size is fine, the dissipative beach type is created.

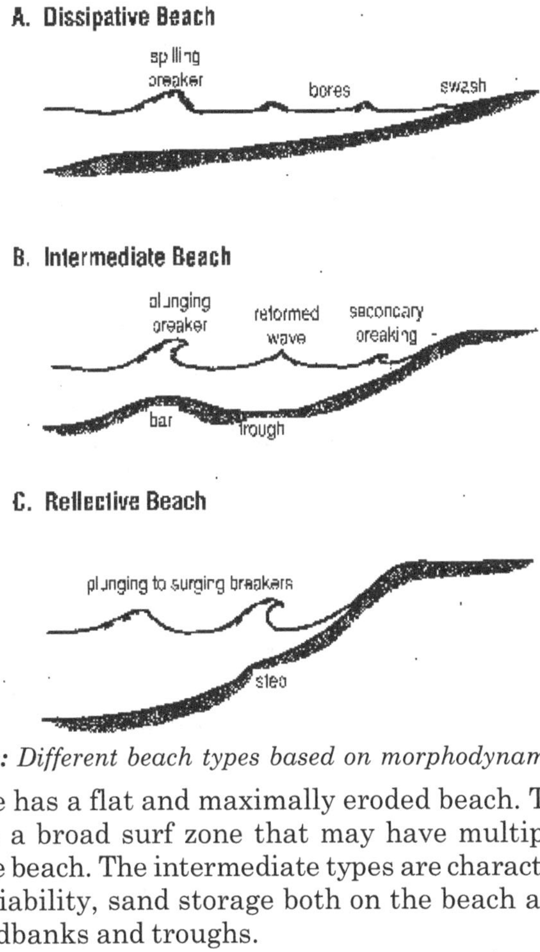

Figure: *Different beach types based on morphodynamic scale*

This type has a flat and maximally eroded beach. The sediments are stored in a broad surf zone that may have multiple sandbanks parallel to the beach. The intermediate types are characterized by high temporal variability, sand storage both on the beach and in the surf zone and sandbanks and troughs.

Beach types can also be based on the degree of exposure. This ranges from very sheltered over sheltered and exposed to very exposed.

Functioning and Adaptations

The intertidal zone is covered part of the day by water and is part of the day exposed to air. High tides bring nutrients and food with it. When the tide retreats, waste products, eggs and larvae are taken. This causes changes for the organisms that live here. They have adapted to this changing environment, as seen on rocky shores.

The burrowing must be rapid and powerful on high-energy sandy beaches. This is because the animals must not be swept away by incoming waves and swash. They also need to be high mobile and must be able to deal with the swash climate. In contrast with rocky shores, desiccation is not an overriding concern, because the animals can retreat into the substratum or below the water table. Intertidal filter-feeders cannot feed while the tide has retreated. Many species of the meiofauna use vertical tidal migrations through the sand column. Other species move up and down the beach with the tides.

This is inadequate for the maintenance of appropriate rhythmic behavior so responses to changing environmental factors are essential. There is a difference between directional (such as light, slope of the beach, water currents) and nondirectional (such as disturbance of the sand, changes in temperature, hydrostatic pressure) stimuli. Directional stimuli act as orientational signs, while nondirectional stimuli act as releasing factors.

Because of the absence of attached macrophytes, the predominant feeding types are filter-feeding and scavenging. Adaptations to respiration of animals in low-energy sandy beaches are different from those on surf-swept beaches. Some adaptations are an increased ventilation rate or increased efficiency, reduced metabolic rate or other ways of conserving energy. Many sheltered-shore animals are facultative anaerobes.

This is an adaptation during ebb tide. Other animals in oxygenated surf-swept beaches are essentially aerobic. The majority of the intertidal animals have tolerance levels of natural variables that exceed those necessary for survival in their particular habitats. Some species descent into the burrow to escape high temperatures.

Another solution is evaporative cooling by replacing water through entering the burrow, plunging into the sea or absorption from the substratum. Another problem for intertidal animals is the time of reproduction. There is variation in the number of eggs, the anatomy of the reproductive organs, the morphology of the egg cases, times of breeding, mating behavior and developmental stages. Adaptations for this is to reproduce at frequent times (iteroparous) or to reproduce just once in a year (semelparous).

This depends from species to species. Some species follow the lunar cycle to reproduce at the right time. To ovoid predation, several behaviors are developed. The first one is to burrow very deep. Another one is tidal migration, so the animals remain protected from predation. Other responses are escaping movements or an impressive threat display by crabs by holding their chelae open and aloft. According to circumstances, the behavior of the animals can be modified. This is called phenotypic plasticity.

Several groups of vertebrates make use of sandy beaches for foraging, nesting and breeding. Turtles nest on the backshore of sandy beaches. Birds use the beach for foraging, nesting and roosting. Seals use several areas of the beach for nesting, molting, breeding and raising pups. Other terrestrial animals such as otters, baboons, raccoons, lions,... They descend onto the beach to forage.

Biota

The distribution and abundance of the sediment infauna is mostly controlled by complex interactions between the physicochemical and biological properties of the sediment.

The physicochemical properties are:

- Grain size
- Water content
- Flushing rate of water through the sediment
- Oxidation-reduction state
- Dissolved oxygen
- Temperature
- Light
- Organic content

The biological properties are:

- Food availability and feeding activity
- Reproductive effects on dispersal and settlement
- Behavior that induces movement and aggregation
- Intraspecific competition
- Interspecific competition and competitive exclusion
- Predation effects

Most invertebrate phyla are represented on sandy beaches, either as interstitial forms or as members of the macrofauna. The macrofaunal forms are by far the better known.

Some of them are typical of intertidal sands and their surf zone, while others are more characteristic of sheltered sandbanks, sandy muds or estuaries and are less common on open beaches of pure sand.

Macrofauna

Macrofauna of the sandy beaches are often abundant and, in some cases, attain exceptionally high densities. Their main feature is the high degree of mobility displayed by all species. These animals may vary from a few mm to 20 cm in length.

The macrofauna community consists of those organisms too large to move between the sand grains. The macrofauna of sandy beaches includes most major invertebrate taxa although it has been recognised that molluscs, crustaceans and polychaetes are the most important.

There is a tendency for crustaceans to be more abundant on tropical sandy beaches or more exposed beaches and molluscs to be more abundant on less exposed and on temperate beaches although there are many exceptions of this and polychaetes are sometimes more abundant than either of these taxa. Generally crustaceans dominate the sands towards the upper tidal level and molluscs the lower down level.

Physical factors, primary wave action and particle size of the sand largely determine distribution and diversity of the invertebrate macrofauna of sandy beaches. Food input and surf-zone productivity may determinate the abundance population. Water movement is important parameter controlling macrofaunal distribution on beaches.

Meiofauna

In contrast to the wave-swept surface sand inhabited by most of the macrofauna, the interstitial system is truly three-dimensional, often having great vertical extent in the sand. The porous system averages about 40% of the total sediment volume. Its inhabitants include small metazoans forming the meiofauna, protozoans, bacteria and diatoms. The meiofauna is defined as those metazoan animals passing undamaged though 0.5 to 1.0 mm sieves and trapped on 30 mm screens. On most beaches the interstitial fauna is rich and diverse, even exceeding the macrofauna in biomass in some cases. The dominant taxa of sandy beach meiofauna are nematodes and harpacticoid copepod with other important groups including turbellarians, oligochaetes, gastrotrichs, ostracods and tardigdades.

Insects

Terrestrial insects and vertebrates are frequently ignored in accounts of sandy beaches. These animals are usually a conspicuous component of the ecosystems, often rivalling the aquatic macrofauna in terms of biomass and having a significant impact on the system with regard to predation and scavenging.

Coastal Dunes

In physical geography, a dune is a hill of sand built by either wind or water flow. Dunes occur in different shapes and sizes, formed by interaction with the flow of air or water. Most kinds of dunes are longer on the windward side where the sand is pushed up the dune and have a shorter "slip face" in the lee of the wind. The valley or trough between dunes is called a *slack*. A "dune field" is an area covered by extensive sand dunes. Large dune fields are known as ergs.

Some coastal areas have one or more sets of dunes running parallel to the shoreline directly inland from the beach. In most cases, the dunes are important in protecting the land against potential ravages by storm waves from the sea. Although the most widely distributed dunes are those associated with coastal regions, the largest complexes of dunes are found inland in dry regions and associated with ancient lake or sea beds.

Dunes can form under the action of water flow (fluvial processes), and on sand or gravel beds of rivers, estuaries and the sea-bed.

The modern word "dune" came into English from French circa 1790. In ancient times, words cognate to "dune" probably had the meaning of a built-up hill or citadel fortification.

Dunes provide a buffer to the coastal system - they protect the regions behind them from wind and waves, and provide a substrate, or ground, for plants to form communities upon. Three things are required for dune formation to occur: a large supply of sand, wind speeds capable of moving it, and an ideal location for its accumulation.

Conservation

Dune habitats provide niches for highly specialised plants and animals, including numerous rare species and some endangered species. Due to widespread human population expansion, dunes face destruction through land development and recreational usages, as well as alteration to prevent the encroachment of sand onto inhabited areas. Some countries, notably the United States, Australia, Canada, New Zealand, the United Kingdom, and Netherlands, have developed significant programs of dune protection through the use of sand dune stabilization. In the U.K., a Biodiversity Action Plan has been developed to assess dunes loss and to prevent future dunes destruction.

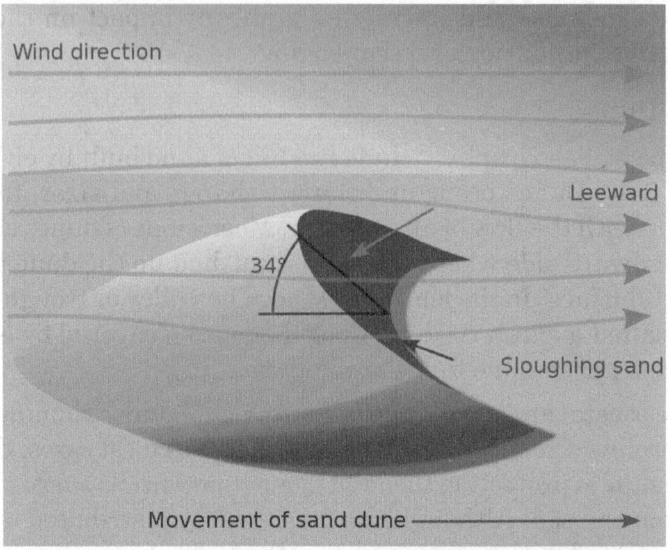

Figure: *Sand blowing off a crest in the Kelso Dunes of the Mojave Desert, California.*

Aeolian Dune Shapes

Crescentic: "Crescent-shaped mounds are generally wider than they are long. The slipfaces are on the concave sides of the dunes. These dunes form under winds that blow consistently from one direction, and they also are known as barchans, or transverse dunes. Some types of crescentic dunes move more quickly over desert surfaces than any other type of dune. A group of dunes moved more than 100 metres per year between 1954 and 1959 in the China's Ningxia Province, and similar speeds have been recorded in the Western Desert of Egypt. The largest crescentic dunes on Earth, with mean crest-to-crest widths of more than 3 kilometres, are in China's Taklamakan Desert."

Linear: "Straight or slightly sinuous sand ridges typically much longer than they are wide are known as linear dunes. They may be more than 160 kilometres (100 miles) long. Some linear dunes merge to form Y-shaped compound dunes. Many form in bidirectional wind regimes. The long axes of these dunes extend in the resultant direction of sand movement."

Linear loess hills known as pahas are superficially similar. These hills appear to have been formed during the last ice age under permafrost conditions dominated by sparse tundra vegetation.

Star: Radially symmetrical, star dunes are pyramidal sand mounds with slipfaces on three or more arms that radiate from the high center of the mound. They tend to accumulate in areas with multidirectional wind regimes. Star dunes grow upward rather than laterally. They dominate the Grand Erg Oriental of the Sahara. In other deserts, they occur around the margins of the sand seas, particularly near topographic barriers. In the southeast Badain Jaran Desert of China, the star dunes are up to 500 metres tall and may be the tallest dunes on Earth.

Dome: Oval or circular mounds that generally lack a slipface. Dome dunes are rare and occur at the far upwind margins of sand seas.

Parabolic: U-shaped mounds of sand with convex noses trailed by elongated arms are parabolic dunes. These dunes are formed from blowout dunes where the erosion of vegetated sand leads to a U-shaped depression. The elongated arms are held in place by vegetation; the largest arm known on Earth reaches 12 km. Sometimes these dunes are called U-shaped, blowout, or hairpin dunes, and they are well known in coastal deserts. Unlike crescent shaped dunes, their crests point upwind. The bulk of the sand in the dune migrates forward.

In plan view, these are U-shaped or V-shaped mounds of well-sorted, very fine to medium sand with elongated arms that extend upwind behind the central part of the dune. There are slip faces that often occur on the outer side of the nose and on the outer slopes of the arms.

These dunes often occur in semiarid areas where the precipitation is retained in the lower parts of the dune and underlying soils. The stability of the dunes was once attributed to the vegetative cover but recent research has pointed to water as the main source of parabolic dune stability. The vegetation that covers them—grasses, shrubs, and trees—help anchor the trailing arms. In inland deserts, parabolic dunes commonly originate and extend downwind from blowouts in sand sheets only partly anchored by vegetation. They can also originate from beach sands and extend inland into vegetated areas in coastal zones and on shores of large lakes.

Most parabolic dunes do not reach heights higher than a few tens of metres except at their nose, where vegetation stops or slows the advance of accumulating sand.

Simple parabolic dunes have only one set of arms that trail upwind, behind the leading nose. Compound parabolic dunes are coalesced features with several sets of trailing arms. Complex parabolic dunes include subsidiary superposed or coalesced forms, usually of barchanoid or linear shapes.

Parabolic dunes, like crescent dunes, occur in areas where very strong winds are mostly unidirectional. Although these dunes are found in areas now characterized by variable wind speeds, the effective winds associated with the growth and migration of both the parabolic and crescent dunes probably are the most consistent in wind direction.

The grain size for these well-sorted, very fine to medium sands is about 0.06 to 0.5 mm. Parabolic dunes have loose sand and steep slopes only on their outer flanks. The inner slopes are mostly well packed and anchored by vegetation, as are the corridors between individual dunes.

Because all dune arms are oriented in the same direction, and, the inter-dune corridors are generally swept clear of loose sand, the corridors can usually be traversed in between the trailing arms of the dune. However to cross straight over the dune by going over the trailing arms, can be very difficult. Also, traversing the nose is very difficult as well because the nose is usually made up of loose sand without much if any vegetation.

A type of extensive parabolic dune that lacks discernible slip faces and has mostly coarse grained sand is known as a Zibar. The term zibar comes from the Arabic word to describe "rolling transverse ridges ... with a hard surface". The dunes are small, have low relief, and can be found in many places across the planet from Wyoming (United States) to Saudi Arabia to Australia. Spacing between zibars ranges from 50 to 400 meters and they don't become more than 10 meters high. The dunes form at about ninety degrees to the prevailing wind which blows away the small, fine-grained sand leaving behind the coarser grained sand to form the crest.

Longitudinal (Seif) Dunes

Longitudinal dunes (also called Seif dunes, after the Arabic word for "sword"), elongate parallel to the prevailing wind, possibly caused by a larger dune having its smaller sides blown away. Seif dunes are sharp-crested and are common in the Sahara. They range up to 300 m (980 ft) in height and 300 km (190 mi) in length. In the southern third of the Arabian Peninsula, a vast erg called the Rub' al Khali or the Empty Quarter, contains seif dunes that stretch for almost 200 km and reach heights of over 300 m.

Seif dunes are thought to develop from barchans if a change of the usual wind direction occurs. The new wind direction will lead to the development of a new wing and the over development of one of the original wings. If the prevailing wind then becomes dominant for a lengthy period of time the dune will revert to its barchan form, with one exaggerated wing. Should the strong wind then return the exaggerated wing will further extend so that eventually it will be supplied with sand when the prevailing wind returns. The wing will continue to grow under both wind conditions, thus producing a seif dune. On a seif dune the slip face develops on the side facing away from the strong wind, while the slip face of a barchan faces the direction of movement. In the sheltered troughs between highly developed seif dunes barchans may be formed because the wind is unidirectional.

A transverse dune is perpendicular to the prevailing wind, probably caused by a steady build-up of sand on an already existing minuscule mound.

Reversing Dunes

Occurring wherever winds periodically reverse direction, reversing dunes are varieties of any of the above shapes. These dunes typically have major and minor slipfaces oriented in opposite directions.

All these dune shapes may occur in three forms: simple, compound, and complex. Simple dunes are basic forms with a minimum number of slipfaces that define the geometric type. Compound dunes are large dunes on which smaller dunes of similar type and slipface orientation are superimposed, and complex dunes are combinations of two or more dune types. A crescentic dune with a star dune superimposed on its crest is the most common complex dune. Simple dunes represent a wind regime that has not changed in intensity or direction since the formation of the dune, while compound and complex dunes suggest that the intensity and direction of the wind has changed.

Sub-aqueous Dunes

Sub-aqueous (underwater) dunes form on a bed of sand or gravel under the actions of water flow. They are ubiquitous in natural channels such as rivers and estuaries, and also form in engineered canals and pipelines. Dunes move downstream as the upstream slope is eroded and the sediment deposited on the downstream or lee slope in typical bedform construction.

These dunes most often form as a continuous 'train' of dunes, showing remarkable similarity in wavelength and height.

Dunes on the bed of a channel significantly increase flow resistance, their presence and growth playing a major part in river flooding.

Lithified Dunes

A lithified (consolidated) sand dune is a type of sandstone that is formed when a marine or aeolian sand dune becomes compacted and hardened. Once in this form, water passing through the rock can carry and deposit minerals, which can alter the color of the rock. Cross-bedded layers of stacks of lithified dunes can produce the cross-hatching patterns, such as those seen in the Zion National Park in the western United States.

A slang term that is used in the Southwestern States (of the U.S.A.) for those consolidated and hardened sand dunes is "slickrock", a name that was introduced by pioneers of the Old West because their steel-rimmed wagon wheels could not gain traction on the rock.

Coastal Dunes

Dunes form where constructive waves encourage the accumulation of sand, and where prevailing onshore winds blow this sand inland. There need to be obstacles—for example, vegetation, pebbles and so on—to trap the moving sand grains. As the sand grains get trapped they start to accumulate, starting dune formation. The wind then starts to affect the mound of sand by eroding sand particles from the windward side and depositing them on the leeward side. Gradually this action causes the dune to "migrate" inland, as it does so it accumulates more and more sand. Dunes provide privacy and shelter from the wind.

Ecological Succession on Coastal Dunes

As a dune forms, plant succession occurs. The conditions on an *embryo dune* are harsh, with salt spray from the sea carried on strong winds. The dune is well drained and often dry, and composed of calcium carbonate from seashells. Rotting seaweed, brought in by storm waves adds nutrients to allow pioneer species to colonize the dune. These pioneer species are marram grass, sea wort grass and other sea grasses in the United Kingdom. These plants are well adapted to the harsh conditions of the foredune typically having deep roots which reach the water table, root nodules that produce nitrogen compounds, and protected stoma, reducing transpiration. Also, the deep roots bind the sand together, and the dune grows into a foredune as more sand is blown over the grasses.

The grasses add nitrogen to the soil, meaning other, less hardy plants can then colonize the dunes. Typically these are heather, heaths and gorses. These too are adapted to the low soil water content and have small, prickly leaves which reduce transpiration. Heather adds humus to the soil and is usually replaced by coniferous trees, which can tolerate low soil pH, caused by the accumulation and decomposition of organic matter with nitrate leaching. Coniferous forests and heathland are common climax communities for sand dune systems.

Young dunes are called yellow dunes and dunes which have high humus content are called grey dunes. Leaching occurs on the dunes, washing humus into the slacks, and the slacks may be much more developed than the exposed tops of the dunes. It is usually in the slacks that more rare species are developed and there is a tendency for the dune slacks soil to be waterlogged and where only marsh plants can survive. These plants would include: creeping willow, cotton grass, yellow ins, reeds, and rushes. As for the species, there is a tendency for natterjack toads to breed here.

Desertification

Sand dunes can have a negative impact on humans when they encroach on human habitats. Sand dunes move via a few different means, all of them helped along by wind. One way that dunes can move is by saltation, where sand particles skip along the ground like a bouncing ball. When these skipping particles land, they may knock into other particles and cause them to move as well, in a process known as creep. With slightly stronger winds, particles collide in mid-air, causing sheet flows. In a major dust storm, dunes may move tens of metres through such sheet flows. Also as in the case of snow, sand avalanches, falling down the slipface of the dunes—that face away from the winds—also move the dunes forward.

Sand threatens buildings and crops in Africa, the Middle East, and China. Drenching sand dunes with oil stops their migration, but this approach is quite destructive to the dunes' animal habitats and uses a valuable resource. Sand fences might also slow their movement to a crawl, but geologists are still analyzing results for the optimum fence designs. Preventing sand dunes from overwhelming towns, villages, and agricultural areas has become a priority for the United Nations Environment Programme. Planting dunes with vegetation also helps to stabilise them.

Sand Supply

Sand comes from many locations and environments, and may be derived from either rock or biotic sources. It can have a source as close as rockfalls from seacliffs, or be transported hundreds to thousands of miles by streams and rivers. A portion is even carried as dust and sand in the air. Once the sand is has been washed into the ocean longshore drift moves and collects it in locations - prompting dune formation.

Sand Type

The majority of the sand you'll see in temperate regions is composed of minerals eroded from bedrock. It is generally a combination of many mineral types and might look something like the sample at right. The primary component of rock-derived sand is silica, what both quartz and glass are made of, and is very resistant to erosion. It remains after other minerals have been broken down mechanically and altered chemically by their environments. Another mineral can also be found in sand - calcium carbonate, or CaCO3, in higher percentages the closer to the equator the sample comes from. Calcium carbonate is a biotic by-product - it is grown as a shell or skeleton by marine organisms suck as molluscs and corals.

Wind

Once sand begins to pile up, ripples and dunes can form. Wind continues to move sand up to the top of the pile until the pile is so steep that it collapses under its own weight. The collapsing sand comes to rest when it reaches just the right steepness to keep the dune stable. This angle, usually about 30-34°, is called the *angle of repose*. Every pile of loose particles has a unique angle of repose, depending upon the properties of the material it's made of, such as the grain size and roundness. Ripples grow into dunes with increase of wind and sand input.

The repeating cycle of sand inching up the windward side to the dune crest, then slipping down the dune's *slip face* allows the dune to inch forward, migrating in the direction the wind blows. As you might guess, all of this climbing then slipping leaves its mark on the internal structure of the dune. The image on the right shows fossil sand dune structure preserved in the Merced Formation at Fort Funston, Golden Gate National Recreation Area. The sloping lines or laminations you see are the preserved slip faces of a migrating sand dune. This structure is called cross-bedding, and can be the result of either wind or water currents. The larger the cross-bedded structure, however, the more likely it is to be formed by wind, rather than water.

A breeze of 10 mi/hr, or 16 km/hr will put fine sand in motion, but will not form dunes. Dune formation relies upon the weight and size of the sand grains in relation to the prevailing, or normal, wind. If the wind blows too hard, any existing dunes will actually be destroyed, as can happen in hurricanes. Once the wind drops back down to normal the dunes start to form again.

Sand Trap

Waves move the sand provided by longshore drift - the migration of sand along the coast by tide and wave action. Waves push the sand up on the beach, and then pull it back out, however the current will cause the overall movement to be in a certain direction. This can be seen in the image at left, the waves are breaking towards the viewer, pushing sand in that direction. When the landscape changes and the longshore drift is disrupted, beaches and dune fields can be cut off from their supply of sand, and shrink. For this reason engineers are very careful of where channels and sea-walls, rip-rap, and groins are placed.

Dune Stability

Because they are entirely composed of loose, or unconsolidated, sand, dunes tend to be very fragile, mobile, and susceptible to deterioration and erosion. They act as a buffer zone, protecting the land behind them from the force of ocean wind and waves. Plants that colonize the dunes act to stabilize the dunes and entrain blowing sediment. These "pioneer" species also act to give other species a foothold, enabling a progression of species that act to further stabilize and enlarge the dune field. Animals, humans and severe weather can all impact negatively on this stabilization. The image to the above right is an experimental plot showing the impact of grazing by horses on dune stability at Assateague National Seashore.

Chapter 5

Coastal Wetlands

A wetland is a land area that is saturated with water, either permanently or seasonally, such that it takes on the characteristics of a distinct ecosystem. Primarily, the factor that distinguishes wetlands from other land forms or water bodies is the characteristic vegetation of aquatic plants, adapted to its unique hydric soil. Wetlands play a number of roles in the environment, principally water purification, flood control, and shoreline stability.

Figure: *Laguna de Rocha, the largest wetland in the urban area in Esteban Echeverría Partido, Argentina*

Wetlands are also considered the most biologically diverse of all ecosystems, serving as home to a wide range of plant and animal life. Wetlands occur naturally on every continent except Antarctica, the largest including the Amazon River basin, the West Siberian Plain, and the Pantanal. The water found in wetlands can be freshwater, brackish, or saltwater.

The main wetland types include swamps, marshes, bogs, and fens; and sub-types include mangrove, carr, pocosin, and varzea. They can also be constructed artificially as a water management tool, which may play a role in the developing field of water-sensitive urban design. The UN Millennium Ecosystem Assessment determined that environmental degradation is more prominent within wetland systems than any other ecosystem on Earth. International conservation efforts are being used in conjunction with the development of rapid assessment tools to inform people about wetland issues.

Types of Wetlands

Almost every country in the world possesses a wetland of some description. Some are seasonally aquatic, some seasonally terrestrial. All play a vital role for humans and nature alike.

Wetlands are highly productive ecosystems that provide the world with nearly two-thirds of its fish harvest.

They take many forms including marshes, estuaries, mudflats, mires, ponds, fens, pocosins, swamps, deltas, coral reefs, billabongs, lagoons, shallow seas, bogs, lakes, and floodplains.

There are also human-made wetlands such as fish and shrimp ponds, farm ponds, irrigated agricultural land, salt pans, reservoirs, gravel pits, sewage farms and canals.

There are many different ways of categorising wetlands, and numerous different types. Most large wetland areas are a unique mosaic of different wetland types. For example, within the The Kafue Flats are found grasslands, lagoons, marshes, swamps and reed beds.

Coastal Wetlands

Coastal wetlands are found in the areas between land and open sea that are not influenced by rivers such as. shorelines, beaches, mangroves and coral reefs.

A good example are the mangrove swamps found in sheltered tropical coastal areas. The partly submerged roots of mangrove trees spread out beneath the water to trap sediment and prevent it being washed out to sea. Around 70% of tropical coastlines are mangrove-lined. Some mangroves are strategically planted between land and sea to stabilise shores and to protect communities from violent storms and powerful waves. The result is an area of stillness and tranquillity where fish breed, wildlife takes refuge, and local people earn a living from occupations as diverse as charcoal burning and bee keeping.

Shallow Lakes and Ponds

These wetlands are areas of permanent or semi-permanent water with little flow. They include vernal ponds, spring pools, salt lakes and volcanic crater lakes. They are small, shallow, intermittently flooded depressions in grasslands or forests, and are often only wet in winter and early spring.

Bogs

Bogs are waterlogged peatlands in old lake basins or depressions in the landscape. Almost all water in bogs comes from rainfall. Bogs have specialised and unique flora that have evolved in their nutrient-poor and acidic conditions, including for example the carnivorous pitcher plant. As bogs are unsuitable for agriculture, forestry or development they offer an undisturbed habitat for a wide range of species, including moose, black bear, lynx, snowshoe hare and mink.

In addition, bogs are used by many species of migratory birds, providing a safe habitat in which to breed, rest and feed. A particular kind of bog is found on the coastal plain of the southeasteern United States. Pocosins are evergreen shrub bogs typically found on high areas of a flat water-logged landscape.

Marshes and Swamps

Also known as palustrine wetlands, marshes, swamps and fens account for almost half of all wetlands throughout the world.

Marshes are one of the broadest categories of wetlands and in general harbour the greatest biological diversity. Marshes form in depressions in the landscape, as fringes around lakes, and along slow-flowing streams and rivers. Marshes are dominated by floating-leafed plants like water lillies and duckweed.

Marshes slow down the rate of rainfall drainage and control its flow into rivers, lakes, and streams.

Estuaries

The area where rivers meet the sea and water changes from fresh to salt can offer an extremely rich mix of biodiversity.

These wetlands include deltas, tidal mudflats and salt marshes. Mudflats and seagrass beds in particular provide a rich diet for many species of insects, birds, fish, turtles and other species.

The Value of Coastal Wetlands for Coping with Sea Level Rise

Coastal wetlands make communities more resilient by providing flood storage, storm surge buffers, erosion control, water quality improvements, and wildlife habitat. Sea level rise is a threat not only to wetlands, but also to the communities these wetlands serve.

Coastal counties generate much of our nation's gross domestic product and support a large percentage of the population. The ability of coastal communities and ecosystems to adapt to sea level rise is critical to the nation's long-term security and economic sustainability. The continuing threat of sea level rise, along with additional climate change-related stressors and growing populations, makes protection of coastal wetlands and the services they provide essential.

Using nature to help people cope with the expected impacts of climate change is a strategy known as ecosystem-based adaptation. To implement this strategy, geospatial data and technology can be used to identify wetlands that are vulnerable and those that have the potential to increase a community's resilience to sea level rise. Armed with this information, communities can prioritize their conservation efforts and spend their conservation capital wisely.

Coastal Wetland Ecosystem Services

Important ecosystem services that coastal wetlands provide include the following:

Flood Storage – Wetlands function as natural sponges that trap and slowly release surface water, rain, snowmelt, groundwater, and flood waters and distribute these waters more slowly over the floodplain, thereby lowering flood heights and dissipating storm surge.

Storm Surge Buffering – Large amounts of wave energy and water from storms that may otherwise do extensive inland damage are absorbed by coastal wetlands. Some research indicates that the height of storm surge can be reduced by one foot for every mile of vegetative wetlands that exists.

Erosion Control – Wetland vegetation stabilizes the shoreline by holding sediments in place with roots, absorbing wave energy, and breaking up the flow of stream or river currents.

Water Quality Maintenance – Wetlands reduce pollution of waterways by storing and filtering urban runoff and removing or retaining nutrients and sediment carried by runoff. Wetlands play a unique and critical function through their ability to recycle these nutrients into usable substances. Wetland functions reduce the costs of constructing, operating, and maintaining drinking water treatment plants.

Fish and Wildlife Habitat – Wetlands are among the most productive ecosystems on earth and contribute to a diversity of plants and animals. Wetlands provide fish and wildlife with breeding, nesting, and nursery grounds, as well as needed shelter, food, and migration stopovers.

Marsh

A marsh is a type of wetland that is dominated by herbaceous rather than woody plant species. Marshes can often be found at the edges of lakes and streams, where they form a transition between the aquatic and terrestrial ecosystems. They are often dominated by grasses, rushes or reeds. If woody plants are present they tend to be low-growing shrubs. This form of vegetation is what differentiates marshes from other types of wetland such as swamps, which are dominated by trees, and mires, which are wetlands that have accumulated deposits of acidic peat.

Marshes provide habitat for many types of plants and animals that have adapted to living in flooded conditions. The plants must be able to survive in wet mud with low oxygen levels. Many of these plants therefore have aerenchyma, channels within the stem that allow air to move from the leaves into the rooting zone. Marsh plants also tend to have rhizomes for underground storage and reproduction. Familiar examples include cattails, sedges, papyrus and sawgrass. Aquatic animals, from fish to salamanders, are generally able to live with a low amount of oxygen in the water. Some can obtain oxygen from the air instead, while others can live indefinitely in conditions of low oxygen. Marshes provide habitat for many kinds of invertebrates, fish, amphibians, waterfowl and aquatic mammals. Marshes have extremely high levels of biological production, some of the highest in the world, and therefore are important in supporting fisheries. Marshes also improve water quality by acting as a sink to filter pollutants and sediment from the water that flows through them. Marshes (and other wetlands) are able to absorb water during periods of heavy rainfall and slowly release it into waterways and therefore reduce the magnitude of flooding. The pH in marshes tends to be neutral to alkaline, as opposed to bogs, where peat accumulates under more acid conditions.

Types of Marshes

Marshes differ depending mainly on their location and salinity. Both of these factors greatly influence the range and scope of animal and plant life that can survive and reproduce in these environments. The three main types of marsh are salt marshes, freshwater tidal marshes, and freshwater marshes. These three can be found worldwide and each contains a different set of organisms.

Salt Marshes

A salt marsh or saltmarsh, also known as a coastal salt marsh or a tidal marsh, is a coastal ecosystem in the upper coastal intertidal zone between land and open salt water or brackish water that is regularly flooded by the tides. It is dominated by dense stands of salt-tolerant plants such as herbs, grasses, or low shrubs. These plants are terrestrial in origin and are essential to the stability of the salt marsh in trapping and binding sediments. Salt marshes play a large role in the aquatic food web and the delivery of nutrients to coastal waters. They also support terrestrial animals and provide coastal protection.

Salt marshes occur on low-energy shorelines in temperate and high-latitudes which can be stable or emerging, or submerging if the sedimentation rate exceeds the subsidence rate. Commonly these shorelines consist of mud or sand flats (known also as tidal flats or abbreviated to mudflats) which are nourished with sediment from inflowing rivers and streams. These typically include sheltered environments such as embankments, estuaries and the leeward side of barrier islands and spits. In the tropics and sub-tropics they are replaced by mangroves; an area that differs from a salt marsh in that instead of herbaceous plants, they are dominated by salt-tolerant trees.

Most salt marshes have a low topography with low elevations but a vast wide area, making them hugely popular for human populations. Salt marshes are located among different landforms based on their physical and geomorphological settings. Such marsh landforms include deltaic marshes, estuarine, back-barrier, open coast, embayments and drowned-valley marshes. Deltaic marshes are associated with large rivers where many occur in Southern Europe such as the Camargue, France in the Rhone delta or the Ebro delta in Spain. They are also extensive within the rivers of the Mississippi Delta in the United States. In New Zealand, most salt marshes occur at the head of estuaries in areas where there is little wave action and high sedimentation. Such marshes are located in Awhitu Regional Park in Auckland, the Manawatu Estuary, and the Avon-Heathcote Estuary in Christchurch.

Back-barrier marshes are sensitive to the reshaping of barriers in the landward side of which they have been formed. They are common along much of the eastern coast of the United States and the Frisian Islands. Large, shallow coastal embayments can hold salt marshes with examples including Morecambe Bay and Portsmouth in Britain and the Bay of Fundy in North America.

Salt marshes are sometimes included in lagoons, and the difference is not very marked; the Venetian Lagoon in Italy, for example, is made up of these sorts of animals and or living organisms belonging to this ecosystem. They have a big impact on the biodiversity of the area.

Formation

Tidal flats gain elevation relative to sea level by sediment accretion, and subsequently the rate and duration of tidal flooding decreases so that vegetation can colonize on the exposed surface. The arrival of propagules of pioneer species such as seeds or rhizome portions are combined with the development of suitable conditions for their germination and establishment in the process of colonisation. When rivers and streams arrive at the low gradient of the tidal flats, the discharge rate reduces and suspended sediment settles onto the tidal flat surface, helped by the backwater effect of the rising tide. Mats of filamentous blue-green algae can fix silt and clay sized sediment particles to their sticky sheaths on contact which can also increase the erosion resistance of the sediments. This assists the process of sediment accretion to allow colonising species (e.g., Salicornia spp.) to grow. These species retain sediment washed in from the rising tide around their stems and leaves and form low muddy mounds which eventually coalesce to form depositional terraces, whose upward growth is aided by a sub-surface root network which binds the sediment. Once vegetation is established on depositional terraces further sediment trapping and accretion can allow rapid upward growth of the marsh surface such that there is an associated rapid decrease in the depth and duration of tidal flooding. As a result, competitive species that prefer higher elevations relative to sea level can inhabit the area and often a succession of plant communities develops.

Tidal Flooding and Vegetation Zonation

Coastal salt marshes can be distinguished from terrestrial habitats by the daily tidal flow that occurs and continuously floods the area. It is an important process in delivering sediments, nutrients and plant water supply to the marsh. At higher elevations in the upper marsh zone, there is much less tidal inflow, resulting in lower salinity levels.

Soil salinity in the lower marsh zone is fairly constant due to everyday annual tidal flow. However, in the upper marsh, variability in salinity is shown as a result of less frequent flooding and climate variations. Rainfall can reduce salinity and evapotranspiration can increase levels during dry periods. As a result, there are microhabitats populated by different species of flora and fauna dependant on their physiological abilities. The flora of a salt marsh is differentiated into levels according to the plants' individual tolerance of salinity and water table levels. Vegetation found at the water must be able to survive high salt concentrations, periodical submersion, and a certain amount of water movement, while plants further inland in the marsh can sometimes experience dry, low-nutrient conditions. It has been found that the upper marsh zones limit species through competition and the lack of habitat protection, while lower marsh zones are determined through the ability of plants to tolerate physiological stresses such as salinity, water submergence and low oxygen levels.

The New England salt marsh is subject to strong tidal influences and shows distinct patterns of zonation. In low marsh areas with high tidal flooding, a monoculture of the smooth cordgrass, *Spartina alterniflora* dominate, then heading landwards, zones of the salt hay, *Spartina patens*, black rush, *Juncus gerardii* and the shrub *Iva frutescens* are seen respectively. These species all have different tolerances that make the different zones along the marsh best suited for each individual.

Plant species diversity is relatively low, since the flora must be tolerant of salt, complete or partial submersion, and anoxic mud substrate. The most common salt marsh plants are glassworts (*Salicornia* spp.) and the cordgrass (*Spartina* spp.), which have worldwide distribution. They are often the first plants to take hold in a mudflat and begin its ecological succession into a salt marsh. Their shoots lift the main flow of the tide above the mud surface while their roots spread into the substrate and stabilize the sticky mud and carry oxygen into it so that other plants can establish themselves as well. Plants such as sea lavenders (*Limonium* spp.), plantains (*Plantago* spp.), and varied sedges and rushes grow once the mud has been vegetated by the pioneer species.

Salt marshes are quite photosynthetically active and are extremely productive habitats. They serve as depositories for a large amount of organic matter and are full of decomposition, which feeds a broad food chain of organisms from bacteria to mammals. Many of the halophytic plants such as cordgrass are not grazed at all by higher animals but die off and decompose to become food for micro-organisms, which in turn become food for fish and birds.

Sediment Trapping, Accretion, and the Role of Tidal Creeks

The factors and processes that influence the rate and spatial distribution of sediment accretion within the salt marsh are numerous. Sediment deposition can occur when marsh species provide a surface for the sediment to adhere to, followed by deposition onto the marsh surface when the sediment flakes off at low tide. The amount of sediment adhering to salt marsh species is dependent on the type of marsh species, the proximity of the species to the sediment supply, the amount of plant biomass, and the elevation of the species.

For example, in a study of the Eastern Chongming Island and Jiuduansha Island tidal marshes at the mouth of the Yangtze River, China, the amount of sediment adhering to the species *Spartina alterniflora*, *Phragmites australis*, and *Scirpus mariqueter* decreased with distance from the highest levels of suspended sediment concentrations (found at the marsh edge bordering tidal creeks or the mudflats); decreased with those species at the highest elevations, which experienced the lowest frequency and depth of tidal inundations; and increased with increasing plant biomass. *Spartina alterniflora*, which had the most sediment adhering to it, may contribute >10% of the total marsh surface sediment accretion by this process.

Salt marsh species also facilitate sediment accretion by decreasing current velocities and encouraging sediment to settle out of suspension. Current velocities can be reduced as the stems of tall marsh species induce hydraulic drag, with the effect of minimising re-suspension of sediment and encouraging deposition.

Measured concentrations of suspended sediment in the water column have been shown to decrease from the open water or tidal creeks adjacent to the marsh edge, to the marsh interior, probably as a result of direct settling to the marsh surface by the influence of the marsh canopy.

Inundation and sediment deposition on the marsh surface is also assisted by tidal creeks which are a common feature of salt marshes. Their typically dendritic and meandering forms provide avenues for the tide to rise and flood the marsh surface, as well as to drain water, and they may facilitate higher amounts of sediment deposition than salt marsh bordering open ocean. Salt marshes do not however require tidal creeks to facilitate sediment flux over their surface although salt marshes with this morphology seem to be rarely studied.

The elevation of marsh species is important; those species at lower elevations experience longer and more frequent tidal floods and therefore have the opportunity for more sediment deposition to occur. Species at higher elevations can benefit from a greater chance of inundation at the highest tides when increased water depths and marsh surface flows can penetrate into the marsh interior.

Human Impacts

The coast is a highly attractive natural feature to humans through its beauty, resources, and accessibility. As of 2002, over half of the world's population was estimated to being living within 60 km of the coastal shoreline, making coastlines highly vulnerable to human impacts from daily activities that put pressure on these surrounding natural environments. In the past, salt marshes were perceived as coastal 'wastelands,' causing considerable loss and change of these ecosystems through land reclamation for agriculture, urban development, salt production and recreation. The indirect effects of human activities such as nitrogen loading also play a major role in the salt marsh area. Salt marshes can suffer from dieback in the high marsh and die-off in the low marsh.

Land Reclamation

Reclamation of land for agriculture by converting marshland to upland was historically a common practice. Dikes were often built to allow for this shift in land change and to provide flood protection further inland. In recent times intertidal flats have also been reclaimed. For centuries, livestock such as sheep and cattle grazed on the highly fertile salt marsh land. Land reclamation for agriculture has resulted in many changes such as shifts in vegetation structure, sedimentation, salinity, water flow, biodiversity loss and high nutrient inputs. There have been many attempts made to eradicate these problems for example, in New Zealand, the cordgrass *Spartina anglica* was introduced from England into the Manawatu River mouth in 1913 to try and reclaim the estuary land for farming. A shift in structure from bare tidal flat to pastureland resulted from increased sedimentation and the cordgrass extended out into other estuaries around New Zealand. Native plants and animals struggled to survive as non-natives out competed them. Efforts are now being made to remove these cordgrass species, as the damages are slowly being recognised.

In the Blyth estuary in Suffolk in eastern England, the mid-estuary reclamations (Angel and Bulcamp marshes) that were abandoned in the 1940s have been replaced by tidal flats with compacted soils from agricultural use overlain with a thin veneer of mud. Little vegetation colonisation has occurred in the last 60–75 years and has been attributed to a combination of surface elevations too low for pioneer species to develop, and poor drainage from the compacted agricultural soils acting as an aquaclude. Terrestrial soils of this nature need to adjust from fresh to saline interstitial water by a change in the chemistry and the structure of the soil, accompanied with fresh deposition of estuarine sediment, before salt marsh vegetation can establish. The vegetation structure, species richness, and plant community composition of salt marshes naturally regenerated on reclaimed agricultural land can be compared to adjacent reference salt marshes to assess the success of marsh regeneration.

Upstream Agriculture

Cultivation of land upstream from the salt marsh can introduce increased silt inputs and raise the rate of primary sediment accretion on the tidal flats, so that pioneer species can spread further onto the flats and grow rapidly upwards out of the level of tidal inundation. As a result, marsh surfaces in this regime may have an extensive cliff at their seaward edge. At the Plum Island estuary, Massachusetts (U.S.A), stratigraphic cores revealed that during the 18th and 19th century the marsh prograded over subtidal and mudflat environments to increase in area from 6 km^2 to 9 km^2 after European settlers deforested the land uptream and increased the rate of sediment supply.

Urban Development and Nitrogen Loading

The conversion of marshland to upland for agriculture has in the past century been overshadowed by conversion for urban development. Coastal cities worldwide have encroached onto former salt marshes and in the U.S. the growth of cities looked to salt marshes for waste disposal sites. Estuarine pollution from organic, inorganic, and toxic substances from urban development or industrialisation is a worldwide problem and the sediment in salt marshes may entrain this pollution with toxic effects on floral and faunal species. Urban development of salt marshes has slowed since about 1970 owing to growing awareness by environmental groups that they provide beneficial ecosystem services. They are highly productive ecosystems, and when net productivity is measured in $g\ m^{-2}\ yr^{-1}$ they are equalled only by tropical rainforests. Additionally, they can help reduce wave erosion on sea walls designed to protect low-lying areas of land from wave erosion.

De-naturalisation of the landward boundaries of salt marshes from urban or industrial encroachment can have negative effects. In the

Avon-Heathcote estuary/Ihutai, New Zealand, species abundance and the physical properties of the surrounding margins were strongly linked, and the majority of salt marsh was found to be living along areas with natural margins in the Avon and Heathcote river outlets; conversely, artificial margins contained little marsh vegetation and restricted landward retreat. The remaining marshes surrounding these urban areas are also under immense pressure from the human population as human-induced nitrogen enrichment enters these habitats. Nitrogen loading through human-use indirectly affects salt marshes causing shifts in vegetation structure and the invasion of non-native species.

Human impacts such as sewage, urban run-off, agricultural and industrial wastes are running into the marshes from nearby sources. Salt marshes are nitrogen limited and with an increasing level of nutrients entering the system from anthropogenic effects, the plant species associated with salt marshes are being restructured through change in competition. For example, the New England salt marsh is experiencing a shift in vegetation structure where *S. alterniflora* is spreading from the lower marsh where it predominately resides up into the upper marsh zone. Additionally, in the same marshes, the reed *Phragmites australis* has been invading the area expanding to lower marshes and becoming a dominant species. *P. australis* is an aggressive halophyte that can invade disturbed areas in large numbers outcompeting native plants. This loss in biodiversity is not only seen in flora assemblages but also in many animals such as insects and birds as their habitat and food resources are altered.

Mosquito Control

Earlier in the 20th century, it was believed that draining salt marshes would help reduce mosquito populations. In many locations, particularly in the northeastern United States, residents and local and state agencies dug straight-lined ditches deep into the marsh flats. The end result, however, was a depletion of killifish habitat. The killifish is a mosquito predator, so the loss of habitat actually led to higher mosquito populations, and adversely affected wading birds that preyed on the killifish. These ditches can still be seen, despite some efforts to refill the ditches.

Freshwater Tidal Marshes

Although considered a freshwater marsh, this form of marsh is affected by the ocean tides. However, without the stresses of salinity at work in its saltwater counterpart, the diversity of the plants and animals that live in and use freshwater tidal marshes is much higher than in salt marshes. The most serious threats to this form of marsh are the increasing size and pollution of the cities surrounding them.

Freshwater Marshes

Ranging greatly in both size and geographic location, freshwater marshes make up the most common form of wetland in North America. They are also the most diverse of the three types of marsh. Some examples of freshwater marsh types in North America are:

Wet Meadows

Wet meadows occur in areas such as shallow lake basins, low-lying depressions, and the land between shallow marshes and upland areas. They also occur on the edges of large lakes and rivers. Wet meadows often have very high plant diversity and high densities of buried seeds. They are regularly flooded but are often dry in the summer.

Vernal Pools

Vernal pools are a type of marsh found only seasonally in shallow depressions in the land. They can be covered in shallow water, but in the summer and fall, they can be completely dry. In western North America, vernal pools tend to form in open grasslands, whereas in the east they often occur in forested landscapes. Further south, vernal pools form in pine savannas and flatwoods. Many amphibian species depend upon vernal pools for spring breeding; these ponds provide habitat free from fish which eat the eggs and young of amphibians. An example is the endangered gopher frog (*Rana sevosa*). Similar temporary ponds occur in other world ecosystems, where they may have local names. However, the term vernal pool can be applied to all such temporary pool ecosystems.

Playa Lakes

Playa lakes are a form of shallow freshwater marsh that occurs in the southern high plains of the United States. Like vernal pools, they are only present at certain times of the year and generally have a circular shape. As the playa dries during the summer, conspicuous plant zonation develops along the shoreline.

Prairie Potholes

Prairie potholes are found in the northern parts of North America. This landscape was once covered by glaciers, and as a result shallow depressions were formed in great numbers. These depressions fill with water in the spring. They provide important breeding habitat for many species of waterfowl. Some pools only occur seasonally while others retain enough water to be present all year.

Riverine Wetlands

Many kinds of marsh occur along the fringes of large rivers. The different types are produced by factors such as water level, nutrients, ice scour, and waves.

Restoration

Some areas of the world have already lost 90% of their wetlands, including marshes. They have been drained to create agricultural land or filled to accommodate urban sprawl. Restoration is the process of returning marshes to the landscape to replace those lost in the past. Restoration can be done on a large scale, such as by allowing rivers to flood naturally in the spring, or on a small scale by returning wetlands to urban landscapes.

Lagoon

A lagoon is a shallow body of water separated from a larger body of water by barrier islands or reefs. Lagoons are commonly divided into coastal lagoons and atoll lagoons. They have also been identified as occurring on mixed-sand and gravel coastlines. There is an overlap between bodies of water classified as coastal lagoons and bodies of water classified as estuaries. Lagoons are common coastal features around the world.

Lagoons are shallow, often elongated bodies of water separated from a larger body of water by a shallow or exposed shoal, coral reef, or similar feature. Some authorities (such as Nybakken) include fresh water bodies in the definition of "lagoon", while others explicitly restrict "lagoon" to bodies of water with some degree of salinity. The distinction between "lagoon" and "estuary" also varies between authorities. Richard A. Davis Jr. restricts "lagoon" to bodies of water with little or no fresh water inflow, and little or no tidal flow, and calls any bay that receives a regular flow of fresh water an "estuary". Davis does state that the terms "lagoon" and "estuary" are "often loosely applied, even in scientific literature." Kusky characterizes lagoons as normally being elongated parallel to the coast, while estuaries are usually drowned river valleys, elongated perpendicular to the coast. When used within the context of a distinctive portion of coral reef ecosystems, the term "lagoon" is synonymous with the term "back reef" or "backreef", which is more commonly used by coral reef scientists to refer to the same area. Coastal lagoons are classified as inland bodies of water. Many lagoons do not include "lagoon" in their common names. Albemarle and Pamlico sounds in North Carolina, Great South Bay between Long Island and the barrier beaches of Fire Island in New York, Isle of Wight Bay, which separates Ocean City, Maryland from the rest of Worcester County, Maryland, Banana River in Florida, Lake Illawarra in New South Wales, Montrose Basin in Scotland, and Broad Water in Wales have all been classified as lagoons, despite their names. In England, The Fleet at Chesil Beach has also been described as a lagoon.

Figure: *Balos coastal lagoon of northwestern Crete. The shallow lagoon is separated from the Mediterranean sea by narrow shoals connecting to a small rocky mountains.*

In Latin America, the term "laguna", which lagoon translates to, is often used to describe a lake, such as Laguna Catemaco. In Portuguese, "lagoa" may be a body of shallow sea water, but also a relatively small freshwater lake not linked to the sea.

Atoll Lagoons

Atoll lagoons form as coral reefs grow upwards while the islands that the reefs surround subside, until eventually only the reefs remain above sea level. Unlike the lagoons that form shoreward of fringing reefs, atoll lagoons often contain some deep (>20m) portions.

Coastal Lagoons

Coastal lagoons form along gently sloping coasts where barrier islands or reefs can develop off-shore, and the sea-level is rising relative to the land along the shore (either because of an intrinsic rise in sea-level, or subsidence of the land along the coast). Coastal lagoons do not form along steep or rocky coasts, or if the range of tides is more than 4 metres (13 ft). Due to the gentle slope of the coast, coastal lagoons are shallow. They are sensitive to changes in sea level due to global warming. A relative drop in sea level may leave a lagoon largely dry, while a rise in sea level may let the sea breach or destroy barrier islands, and leave reefs too deep under water to protect the lagoon. Nybakken describes coastal lagoons and barrier islands as a "coupled system". Coastal lagoons are young and dynamic, and may be short-lived in geological terms. Coastal lagoons are common, occurring along nearly 15 percent of the world's shorelines. In the United States, lagoons are found along more than 75 percent of the eastern and Gulf coasts.

Coastal lagoons are usually connected to the open ocean by inlets between barrier islands. The number and size of the inlets, precipitation, evaporation, and inflow of fresh water all affect the nature of the lagoon. Lagoons with little or no interchange with the open ocean, little or no inflow of fresh water, and high evaporation rates, such as Lake St. Lucia, in South Africa, may become highly saline. Lagoons with no connection to the open ocean and significant inflow of fresh water, such as the Lake Worth Lagoon in Florida in the middle of the 19th century, may be entirely fresh. On the other hand, lagoons with many wide inlets, such as the Wadden Sea, have strong tidal currents and mixing. Coastal lagoons tend to accumulate sediments from inflowing rivers, from runoff from the shores of the lagoon, and from sediment carried into the lagoon through inlets by the tide. Large quantities of sediment may be occasionally be deposited in a lagoon when storm waves overwash barrier islands. Mangroves and marsh plants can facilitate the accumulation of sediment in a lagoon. Benthic organisms may stabilize or destabilize sediments.

River-mouth Lagoons on Mixed Sand and Gravel Beaches

River-mouth lagoons on mixed sand and gravel (MSG) beaches form at the river-coast interface where a typically braided, although sometimes meandering, river interacts with a coastal environment that is significantly affected by longshore drift. The lagoons which form on the MSG coastlines are common on the east coast of the South Island, New Zealand and have long been referred to as hapua by the Mâori. This classification differentiates hapua from similar lagoons located on the New Zealand coast termed waituna. Hapua are often located on paraglacial coastal areas where there is a low level of coastal development and minimal population density. Hapua form as the river carves out an elongated coast-parallel area, blocked from the sea by a MSG barrier which constantly alters its shape and volume due to longshore drift. Longshore drift continually extends the barrier behind which the hapua forms by transporting sediment along the coast. Hapua are defined as a narrow shore-parallel extensions of the coastal riverbed. They discharge the majority of stored water to the ocean via an ephemeral and highly mobile drainage channel or outlet. The remainder percolates through the MSG barrier due to its high levels of permeability. Hapua systems are driven by a wide range of dynamic processes that are generally classified as fluvial or marine; changes in the balance between these processes as well as the antecedent barrier conditions can cause shifts in the morphology of the hapua, in particular the barrier. New Zealand examples include the Rakaia, Ashburton and Hurunui river-mouths.

Hapua Environment

Hapua have been identified as establishing in the Canterbury Bight coastal region on the east coast of the South Island. They are often found in areas of coarse-grained sediment where contributing rivers have moderately steep bed gradients. MSG beaches in the Canterbury Bight region contain a wide range of sediment sizes from sand to boulders and are exposed to the high energy waves that make up an east coast swell environment. MSG beaches are reflective rather than dissipative energy zones due to their morphological characteristics. They have a steep foreshore which is known as the 'engine room' of the beach profile. In this zone, swash and backwash are dominating processes alongside longshore transport. MSG beaches do not have a surf zone; instead a single line of breakers is visible in all sea conditions. Hapua are associated with MSG beaches as the variation in sediment size allows for the barrier to be permeable.

The east coast of the South Island has been identified as being in a period of chronic erosion of approximately 0.5 metres per year. This erosion trend is a result of a number of factors. According to the classification scheme of Zenkovich, the rivers on the east coast can be described as 'small'; this classification is not related to their flow rate but to the insufficient amount of sediment that they transport to the coast to nourish it. The sediment provided is not adequate to nourish the coast against its typical high energy waves and strong longshore drift. These two processes constantly remove sediment depositing it either offshore or further up drift. As the coastline becomes eroded the hapua have been 'rolling back' by eroding the backshore to move landwards.

Hapua or river-mouth lagoons form in micro-tidal environments. A micro-tidal environment is where the tidal range (distance between low tide and high tide) is less than two metres. Tidal currents in a micro-tidal zone are less than those found on meso-tidal (two – four metres) and macro-tidal (greater than four metres) coastlines. Hapua form in this type of tidal environment as the tidal currents are unable to compete with the powerful freshwater flows of the rivers therefore there is no negligible tidal penetration to the lagoon. A fourth element of the environment in which hapua form is the strong longshore drift component. Longshore or littoral drift is the transportation of sediments along the coast at an angle to the shoreline. In the Canterbury Bight coastal area; the dominant swell direction is northwards from the Southern Ocean. Therefore the principal movement of sediment via longshore drift is north towards Banks Peninsula. Hapua are located in areas dominated by longshore drift; because it aids the formation of the barrier behind which the hapua is sited.

A hapua also requires sediment to form the lagoon barrier. Sediment which nourishes the east coast of New Zealand can be sourced from three different areas. Material from the highly erodible Southern Alps is removed via weathering; then carried across the Canterbury Plains by various braided rivers to the east coast beaches. The second source of sediment is the high cliffs which are located in the hinterland of lagoons. These can be eroded during the occurrence of high river flow or sea storm events. Beaches further south provide nourishment to the northern coast via longshore transport.

Hapua Characteristics

Hapua have a number of characteristics which includes shifts between a variety of morphodynamic states due to changes in the balance between marine and fluvial processes as well as the antecedent barrier conditions. The MSG barrier constantly changes size and shape as a result of the longshore drift. Water stored in the hapua drains to the coast predominately though an outlet; although it can also seep through the barrier depending on the permeability of the material.

Changes in the level of the lagoon water do not occur as a result of saltwater or tidal intrusion. Water in a hapua is predominately freshwater originating from the associated river. Hapua are non-estuarine, there is no tidal inflow however the tide does have an effect on the level of water in the lagoon. As the tide reaches its peak, the lagoon water has a much smaller amount of barrier to permeate through so the lagoon level rises. This is related to a physics theory known as hydraulic head. The lagoon level has a similar sinusoidal wave shape as the tide but reaches its peak slightly later. In general, any saltwater intrusion into the hapua will only occur during a storm via wave overtopping or sea spray.

Hapua can act as both a source and sink of sediment. The majority of sediment in the hapua is fluvial sourced. During medium to low river flows, coarser sediment generally collects in the hapua; while some of the finer sediment can be transported through the outlet to the coast. During flood events the hapua is 'flushed out' with larger amounts of sediment transferred through the outlet. This sediment can be deposited offshore or downdrift of the hapua replenishing the undernourished beach. If a large amount of material is released to the coast at one time it can be identified as a 'slug'. These can often be visible from aerial photographs.

Antecedent barrier conditions combined with changes in the balance between marine and fluvial processes results in shifts between a variety of morphological states in a hapua or river-mouth lagoon on a MSG beach. Marine processes includes the direction of wave approach, wave height and the coincidence of storm waves with high tides. Marine processes tend to dominate the majority of morphodynamic conditions until there is a large enough flood event in the associated river to breach the barrier. The level and frequency of base or flood flows are attributed to fluvial processes. Antecedent barrier conditions are the permeability, volume and height of the barrier as well as the width and presence of previous outlet channels. During low to medium river flows, the outlet from the lagoon to the sea becomes offset in the direction of longshore drift. Outlet efficiency tends to decrease the further away from the main river-mouth the outlet is. A decrease in efficiency can cause the outlet to become choked with sediment and the hapua to close temporarily. The potential for closure varies between different hapua depending on whether marine or fluvial processes are the bigger driver in the event. A high flow event; such as a fresh or flood can breach the barrier directly opposite the main river channel. This causes an immediate decrease in the water level of the hapua; as well as transporting previously deposited sediments into the ocean. Flood events are important for eroding lagoon back shores; this is a behaviour which allows hapua to retreat landward and thus remain coastal landforms even with coastal transgression and sea level rise. During high flow events there is also the possibility for secondary breaches of the barrier or lagoon truncation to occur.

Storm events also have the ability to close hapua outlets as waves overtop the barrier depositing sediment and choking the scoured channel. The resultant swift increase in lagoon water level causes a new outlet to be breached rapidly due to the large hydraulic head that forms between the lagoon and sea water levels. Storm breaching is believed to be an important but unpredictable control on the duration of closures at low to moderate river flow levels in smaller hapua.

Hapua are extremely important for a number of reasons. They provide a link between the river and sea for migrating fish as well as a corridor for migratory birds. To lose this link via closure of the hapua outlet could result in losing entire generations of specific species as they may need to migrate to the ocean or the river as a vital part of their lifecycle. River-mouth lagoons such as hapua were also used a source for *mahinga kai* (food gathering) by the Mâori people. However, this is no longer the case due to catchment degradation which has resulted in lagoon deterioration. River-mouth lagoons on MSG beaches are not well explained in international literature.

Hapua Case Study

The hapua located at the mouth of the Rakaia River stretches approximately three kilometres north from where the river-mouth reaches the coast. The average width of the hapua between 1952 and 2004 was approximately 50 metres; whilst the surface area has stabilised at approximately 600,000 square metres since 1966. The coastal hinterland is composed of erodible cliffs and a low lying area commonly known as the Rakaia Huts. This area has changed notably since European Settlement; with the drainage of ecologically significant wetlands and development of the small bach community.

The Rakaia River begins in the Southern Alps, providing approximately 4.2 Mt per year of sediment to the east coast. It is a braided river with a catchment area of 3105 kilometres squared and a mean flow of 221 cubic metres per second. The mouth of the Rakaia River reaches the coast south of Banks Peninsula. As the river reaches the coast it diverges into two channels; with the main channel flowing to the south of the island. As the hapua is located in the Canterbury Bight it is in a state of constant morphological change due to the prevailing southerly sea swells and resultant northwards longshore drift.

Mangrove

Mangroves are various large and extensive types of trees up to medium height and shrubs that grow in saline coastal sediment habitats in the tropics and subtropics – mainly between latitudes 25° N and 25° S. The remaining mangrove forest areas of the world in 2000 was

53,190 square miles (137,760 km^2) spanning 118 countries and territories.

***Figure:** Mangroves in Malaysia*

The word is used in at least three senses: (1) most broadly to refer to the habitat and entire plant assemblage or mangal, for which the terms mangrove forest biome, mangrove swamp and mangrove forest are also used, (2) to refer to all trees and large shrubs in the mangrove swamp, and (3) narrowly to refer to the mangrove family of plants, the Rhizophoraceae, or even more specifically just to mangrove trees of the genus *Rhizophora*.

The mangrove biome, or mangal, is a distinct saline woodland or shrubland habitat characterized by depositional coastal environments, where fine sediments (often with high organic content) collect in areas protected from high-energy wave action. The saline conditions tolerated by various mangrove species range from brackish water, through pure seawater (30 to 40 ppt(parts per thousand)), to water concentrated by evaporation to over twice the salinity of ocean seawater (up to 90 ppt).

Ecology

Mangrove swamps are found in tropical and subtropical tidal areas. Areas where mangal occurs include estuaries and marine shorelines.

The intertidal existence to which these trees are adapted represents the major limitation to the number of species able to thrive in their habitat. High tide brings in salt water, and when the tide recedes, solar evaporation of the seawater in the soil leads to further increases in salinity. The return of tide can flush out these soils, bringing them back to salinity levels comparable to that of seawater. At low tide, organisms are also exposed to increases in temperature and desiccation, and are then cooled and flooded by the tide. Thus, for a plant to survive in this environment, it must tolerate broad ranges of salinity, temperature, and moisture, as well as a number of other key environmental factors — thus only a select few species make up the mangrove tree community.

About 110 species are considered "mangroves", in the sense of being a tree that grows in such a saline swamp, though only a few are from the mangrove plant genus, *Rhizophora.* However, a given mangrove swamp typically features only a small number of tree species. It is not uncommon for a mangrove forest in the Caribbean to feature only three or four tree species. For comparison, the tropical rainforest biome contains thousands of tree species, but this is not to say mangrove forests lack diversity. Though the trees themselves are few in species, the ecosystem these trees create provides a home for a great variety of other organisms.

Mangrove plants require a number of physiological adaptations to overcome the problems of anoxia, high salinity and frequent tidal inundation. Each species has its own solutions to these problems; this may be the primary reason why, on some shorelines, mangrove tree species show distinct zonation. Small environmental variations within a mangal may lead to greatly differing methods for coping with the environment. Therefore, the mix of species is partly determined by the tolerances of individual species to physical conditions, such as tidal inundation and salinity, but may also be influenced by other factors, such as predation of plant seedlings by crabs.

Once established, mangrove roots provide an oyster habitat and slow water flow, thereby enhancing sediment deposition in areas where it is already occurring. The fine, anoxic sediments under mangroves act as sinks for a variety of heavy (trace) metals which colloidal particles in the sediments scavenged from the water. Mangrove removal disturbs these underlying sediments, often creating problems of trace metal contamination of seawater and biota.

Mangrove swamps protect coastal areas from erosion, storm surge (especially during hurricanes), and tsunamis. The mangroves' massive root systems are efficient at dissipating wave energy. Likewise, they slow down tidal water enough so its sediment is deposited as the tide comes in, leaving all except fine particles when the tide ebbs. In this way, mangroves build their own environments. Because of the uniqueness of mangrove ecosystems and the protection against erosion they provide, they are often the object of conservation programs, including national biodiversity action plans.

However, mangrove swamps' protective value is sometimes overstated. Wave energy is typically low in areas where mangroves grow, so their effect on erosion can only be measured over long periods. Their capacity to limit high-energy wave erosion is limited to events such as storm surges and tsunamis. Erosion often occurs on the outer sides of bends in river channels that wind through mangroves, while new stands of mangroves are appearing on the inner sides where sediment is accruing.

The unique ecosystem found in the intricate mesh of mangrove roots offers a quiet marine region for young organisms. In areas where roots are permanently submerged, the organisms they host include algae, barnacles, oysters, sponges, and bryozoans, which all require a hard surface for anchoring while they filter feed. Shrimps and mud lobsters use the muddy bottoms as their home. Mangrove crabs munch on the mangrove leaves, adding nutritients to the mangal muds for other bottom feeders. In at least some cases, export of carbon fixed in mangroves is important in coastal food webs.

Mangrove plantations in Vietnam, Thailand, the Philippines and India host several commercially important species of fishes and crustaceans. Despite restoration efforts, developers and others have removed over half of the world's mangroves in recent times.

Biology

Of the recognized 110 mangrove species, only about 54 species in 20 genera from 16 families constitute the "true mangroves", species that occur almost exclusively in mangrove habitats. Demonstrating convergent evolution, many of these species found similar solutions to the tropical conditions of variable salinity, tidal range (inundation), anaerobic soils and intense sunlight. Plant biodiversity is generally low in a given mangal. The greatest biodiversity occurs in the mangal of New Guinea, Indonesia and Malaysia.

Adaptations to Low Oxygen

Red mangroves, which can survive in the most inundated areas, prop themselves above the water level with stilt roots and can then absorb air through pores in their bark (lenticels). Black mangroves live on higher ground and make many pneumatophores (specialised root-like structures which stick up out of the soil like straws for breathing) which are also covered in lenticels.

Figure: *A red mangrove, Rhizophora mangle.*

These "breathing tubes" typically reach heights of up to 30 cm, and in some species, over 3 m. The four types of pneumatophores are stilt or prop type, snorkel or peg type, knee type, and ribbon or plank type. Knee and ribbon types may be combined with buttress roots at the base of the tree. The roots also contain wide aerenchyma to facilitate transport within the plants.

Limiting Salt Intake

Red mangroves exclude salt by having significantly impermeable roots which are highly suberised, acting as an ultrafiltration mechanism to exclude sodium salts from the rest of the plant. Analysis of water inside mangroves has shown 90% to 97% of salt has been excluded at the roots. In a frequently cited concept that has become known as the "sacrificial leaf", salt which does accumulate in the shoot then concentrates in old leaves, which the plant then sheds. However, recent research suggests the older, yellowing leaves have no more measurable salt content than the other, greener leaves. Red mangroves can also store salt in cell vacuoles. As seen in the picture on the right, white (or grey) mangroves can secrete salts directly; they have two salt glands at each leaf base (correlating with their name—they are covered in white salt crystals).

Limiting Water Loss

Because of the limited fresh water available in salty intertidal soils, mangroves limit the amount of water they lose through their leaves. They can restrict the opening of their stomata (pores on the leaf surfaces, which exchange carbon dioxide gas and water vapour during photosynthesis). They also vary the orientation of their leaves to avoid the harsh midday sun and so reduce evaporation from the leaves. Anthony Calfo, a noted aquarium author, observed anecdotally a red mangrove in captivity only grows if its leaves are misted with fresh water several times a week, simulating the frequent tropical rainstorms.

Nutrient Uptake

Because the soil is perpetually waterlogged, little free oxygen is available. Anaerobic bacteria liberate nitrogen gas, soluble iron, inorganic phosphates, sulfides, and methane, which make the soil much less nutritious. Pneumatophores (aerial roots) allow mangroves to absorb gases directly from the atmosphere, and other nutrients such as iron, from the inhospitable soil. Mangroves store gases directly inside the roots, processing them even when the roots are submerged during high tide.

Increasing Survival of Offspring

In this harsh environment, mangroves have evolved a special mechanism to help their offspring survive. Mangrove seeds are buoyant and are therefore suited to water dispersal. Unlike most plants, whose seeds germinate in soil, many mangroves (e.g. red mangrove) are viviparous, whose seeds germinate while still attached to the parent tree. Once germinated, the seedling grows either within the fruit (e.g. *Aegialitis*, *Avicennia* and *Aegiceras*), or out through the fruit (e.g. *Rhizophora*, *Ceriops*, *Bruguiera* and *Nypa*) to form a propagule (a ready-to-go seedling) which can produce its own food via photosynthesis.

The mature propagule then drops into the water, which can transport it great distances. Propagules can survive desiccation and remain dormant for over a year before arriving in a suitable environment. Once a propagule is ready to root, its density changes so the elongated shape now floats vertically rather than horizontally. In this position, it is more likely to lodge in the mud and root. If it does not root, it can alter its density and drift again in search of more favorable conditions.

Taxonomy and Evolution

The following listing (modified from Tomlinson, 1986) gives the number of species of mangroves in each listed plant genus and family. Mangrove environments in the Eastern Hemisphere harbor six times as many species of trees and shrubs as do mangroves in the New World. Genetic divergence of mangrove lineages from terrestrial relatives, in combination with fossil evidence, suggests mangrove diversity is limited by evolutionary transition into the stressful marine environment, and the number of mangrove lineages has increased steadily over the Tertiary with little global extinction.

Geographical Regions

Mangroves can be found in over 118 countries and territories in the tropical and subtropical regions of the world. The largest percentage of mangroves is found between the 5° N and 5° S latitudes. Approximately 75% of world's mangroves are found in just 15 countries. Asia has the largest amount (42%) of the world's mangroves, followed by Africa (21%), North/Central America (15%), Oceania (12%) and South America (11%).

Africa

There are important mangrove swamps in Kenya, Tanzania, République Démocratique du Congo (RDC) and Madagascar, with the latter even admixing at the coastal verge with dry deciduous forests.

Nigeria has Africa's largest mangrove concentration, spanning 36,000 km^2. Oil spills and leaks have destroyed many in the last 50 years, damaging the local fishing economy and water quality.

Along the coast of the Red Sea, both on the Egyptian side and in the Gulf of Aqaba, mangroves composed primarily of *Avicennia marina* and *Rhizophora mucronata* grow in about 28 stands that cover about 525 hectares. Almost all Egyptian mangrove stands are now protected.

Americas

Mangroves live in many parts of the tropical and subtropical coastal zones of North, South and Central America.

Continental United States

Because of their sensitivity to subfreezing temperatures, mangroves in the continental United States are limited to the Florida peninsula and isolated growths of black mangrove (*Avicennia germinans*) along the coast of southern Louisiana and South Texas.

Central America and Caribbean

Mangroves occur on the west coast of Costa Rica, on the Pacific and Caribbean coasts of Nicaragua, Belize, Guatemala, Honduras, and Panama, and on many Caribbean Islands, such as Aruba, Curaçao,

Bonaire, Antigua, Anguilla, the Bahamas, Saint Kitts and Nevis, St. Lucia, Virgin Islands and the San Bernardo islands in Colombia. Significant mangals include the Marismas Nacionales-San Blas mangroves in Mexico. Mangroves can also be found in Puerto Rico, Guyana, Cuba, the Dominican Republic, Haiti, Jamaica, Cayman Islands, Trinidad, Barbados, and the Pacific coast of El Salvador.

Belize

The nation of Belize has the highest overall percentage of forest cover of any of the Central American countries. In terms of Belize's mangrove cover - which assumes the form not only of mangrove 'forest', but also of scrubs and savannas, among others - a 2010 satellite-based study of Belize's mangroves by the World Wildlife Fund (WWF) and the Water Center for the Humid Tropics of Latin America and the Caribbean found, in 2010, mangroves covered some 184,548 acres (74,684 hectares) or 3.4% of Belize's territory.

In 1980, by contrast, mangrove cover stood at 188,417 acres (76,250 hectares) - also 3.4% of Belize's territory, although based on the work of mangrove researcher Simon Zisman, Belize's mangrove cover in 1980 was estimated to represent 98.7% of the precolonial extent of those ecosystems. Belize's mangrove cover in 2010 was thus estimated to represent 96.7% of the precolonial cover. Assessing changes in Belize's mangrove cover over a 30-year period was possible because of Belize's participation in the Regional Visualization and Monitoring System, a regional observatory jointly implemented by CATHALAC, RCMRD, ICIMOD, NASA, USAID, and other partners.

South America

Brazil contains approximately 26,000 km^2 of mangals, 15% of the world's total.

Ecuador has substantial remaining mangrove forests in the provinces of El Oro, Guayas, Manabi and Esmeraldas with limited forest remaining in Santa Elena. The northern portion of Esmeraldas province has a large pristine mangrove forest that is preserved as the Reserva Ecológica Cayapas-Mataje (REMACAN) and is an original Ramsar site. This forest is the most preserved within Ecuador and likely the most pristine forest along the Pacific Coast of the Americas.

The only other major mangrove holding in Esmeraldas is in-and-around the community of Muisne and the Rio Muisne Estuary Swampland Wildlife Refuges. The mangroves in-and-around the estuaries of Muisne have decreased in area from 3222 ha in 1971 to

1065 ha as of 2005, during this time commercial shrimp aquaculture has become the dominant land-cover within this estuary environment.

On the border of Esmeraldas province and Manabv province is a formerly large area of mangrove within Cojimies Estuary. The mangroves in this estuary are some of the most degraded in Ecuador with only 19% of 1971 mangrove area remaining as of 1998, although mangrove has recovered since this date. Within Manabv the major mangrove holding estuary is the Chone estuary situated near the city of Bahía de Caráquez. Again, Chone has undergone substantial mangrove deforestation since the advent of commercial aquaculture in Ecuador. Although mangrove loss appears to have halted in this estuary and mangrove regrowth driven by local fisherman is now occurring.

Peru has a very small region of mangrove located in the north-west of the country on the Ecuadorian Border.

Venezuela's northern Caribbean island, Margarita, possesses mangrove forests in the Parque nacional Laguna de La Restinga. Venezuela has 4% of the world's mangroves, with an extension of 6735;km^2.

Colombia possesses large mangrove forests on both its Caribbean and Pacific coasts.

Asia

Indomalaya Ecozone: Mangroves occur on Asia's south coast, throughout the Indian subcontinent, in all Southeast Asian countries, and on islands in the Indian Ocean, Arabian Sea, Bay of Bengal, South China Sea and the Pacific.

The mangal is particularly prevalent in the deltas of large Asian rivers. The Sundarbans is the largest mangrove forest in the world, located in the Ganges River delta in Bangladesh and West Bengal, India.

The Pichavaram mangroves in Tamil Nadu is India's largest mangrove forest and the second largest in the world, The Bhitarkanika Mangroves Forest of Odisha, by the Bay of Bengal, is India's second largest mangrove forest. Other major mangals occur on the Andaman and Nicobar Islands and the Gulf of Kutch in Gujarat.

Mangroves occur in certain muddy swampy islands of the Maldives.

On the Malayan Peninsula mangroves cover an estimated 1,089.7 square kilometres (420.7 sq mi), while most of the remaining 5,320 square kilometres (2,054 sq mi) mangroves in Malaysia are on the island of Borneo. In Vietnam, mangrove forests grow along the southern coast, including two forests: the Can Gio Mangrove Forest biosphere reserve and the U Minh mangrove forest in the sea and coastal region of Kiên Giang, Cà Mau and B¡c Liêu provinces.

The mangrove forests of Kompong Sammaki in Cambodia are of major ecological and cultural importance, as the human population relies heavily on the crabs and fish that live in the roots.

The three most important mangrove forests of Taiwan are: Tamsui River in Taipei, Jhonggang River in Miaoli and the Sihcao Wetlands in Tainan. According to research, four types of mangrove exist in Taiwan. Some places have been developed as scenic areas, such as the log raft routes in Sihcao.

Indonesia

In the Indonesian Archipelago, mangroves occur around much of Sumatra, Borneo, Sulawesi, and the surrounding islands, while further north, they are found along the coast of the Malay Peninsula. Indonesia has around 9.36 million hectares of mangrove forests, but 48% is categorized as 'moderately damaged' and 23% as 'badly damaged'.

Pakistan

Pakistani mangroves are located mainly along the delta of the Indus River (the Indus River Delta-Arabian Sea mangroves ecoregion). Major mangrove forests are found on the coastline of the provinces of Sindh and Balochistan. In Karachi, land reclamation projects have led to the cutting down of mangrove forests for commercial and urban development. On 22 June 2013, 750,000 mangrove saplings were planted at Kharo Chan, Thatta, in a little over 12 hours. This is the highest number of saplings planted within a day.

Middle East

Oman, near Muscat, supports large areas of mangroves, in particular at Shinas, Qurm Park and Mahout Island. In Arabic, mangrove trees are known as *qurm*, thus the mangrove area in Oman is known as Qurm Park. A small mangrove area is present in the Kingdom of Bahrain. Mangroves are also present extensively in neighboring Yemen.

Iranian mangrove forests occur between 25°112 N to 27°522 N. These forests exist in the north part of the Persian Gulf and Sea of Oman, along three maritime provinces in the south of Iran. These provinces, respectively, from southwest to southeast of Iran, include Bushehr, Hormozgan, and Sistan and Balouchestan.

Mangrove is also widely seen in Tarut Island, east of Qatif in Saudi Arabia. In addition, large forest of mangrove surround the coast to the south of Qatif (Siahat Beach). Nonetheless, because of sea land reclaiming the mangrove is being cut down which makes lots of sea fish losses their natural habitats.

The mangrove forests that cover thousands of hectares of land along the UAE shoreline form an integral part of its coastal ecosystem. The Environment Agency – Abu Dhabi (EAD) is currently working on rehabilitation, conservation and protection of mangrove forests in seven key sites in Abu Dhabi including: Saadiyat Island, Jubail Island, Marawah Marine Biosphere Reserve (which also comprises famous Bu Tinah Island), Bu Syayeef Protected Area, Ras Gharab, the Eastern Corniche and Ras Ghanada.

India

The Mangrove tidal forests are found in the areas of coasts influenced by tides. Mud and silt get accumulated on such coasts. Dense Mangroves are common varieties with roots of plants submerged under water. The deltas of Ganga, the Mahanandi, the Krishna, the Godavari, and the Kaveri are covered by such vegetation.

In the Ganga-Bramhaputra delta, Sundri trees are found, which provide durable hard timber. Palm, Coconut, keora, agar, also grow in some parts of the delta. Royal Bengal Tiger is the famous animal in these forests. Turtles Crocodiles, Gharials and Snakes are also found in these forests.

Bangladesh

The Sundarbans is the largest single block of tidal halophytic mangrove forest in the world.[1] The Sunderbans is a UNESCO World Heritage Site covering parts of Khulna Division of Bangladesh.

The Sundarbans National Park is a National Park, Tiger Reserve, and a Biosphere Reserve. It is in the Sundarbans delta in the Indian state of West Bengal. This region is covered by mangrove forests. It is one of the largest reserves for the Bengal tiger. Sundarbans mangrove forest is the largest in the world, and covers areas of India and Bangladesh for more than 80 kilometers in forming Sundarbans National Park, declared a World Heritage Site by Unesco. A third of this area is covered by water and marshes, as well Sundarbans since 1966 has been considered a sanctuary for wildlife because it is estimated that there live about 400 Royal Bengal Tigers and more than 30,000 deer in this area. The forest lies at the feet of the Ganges and is spread across areas of Bangladesh and West Bengal, India, forming the seaward fringe of the delta. The seasonally-flooded Sundarbans freshwater swamp forests lie inland from the mangrove forests. The forest covers 10,000 km^2 of which about 6,000 are in Bangladesh. It became inscribed as a UNESCO world heritage site in 1997, but while the Bangladeshi and Indian portions constitute the same continuous ecotope, these are separately listed in the UNESCO world heritage list as the Sundarbans and Sundarbans National Park, respectively. The Sundarbans is intersected by a complex network of tidal waterways, mudflats and small islands of salt-tolerant mangrove forests. The area is known for the eponymous Royal Bengal Tiger (Panthera tigris tigris), as well as numerous fauna including species of birds, spotted deer, crocodiles and snakes. It is estimated that there are now 500[citation needed] Bengal tigers and about 30,000 spotted deer in the area. Sundarbans was designated a Ramsar site on May 21, 1992. The fertile soils of the delta have been subject to intensive human use for centuries, and the ecoregion has been mostly converted to intensive agriculture, with few enclaves of forest remaining. The remaining forests, together with the Sundarbans mangroves, are important habitat for the endangered tiger. Additionally, the Sundarbans serves a crucial function as a protective flood barrier for the millions of inhabitants in and around Kolkata (Calcutta) against the result of cyclone activity. Sundarbans is home to many different species of birds, mammals, insects, reptiles and fish. It is estimated that there may be found more than 120 species of fish and over 260 species of birds and more than fifty species of reptiles and eight amphibians. Many tourists go there to see the Bengal tigers, saltwater crocodiles, leopards and snakes cobra. Geographical data of Sundarban Latitude: 25.7667 Longitude: 88.7167 Average Height: 34 Time Zone: Asia / Dhaka Title: Lugar Poblado

Oceania

Australia and New Guinea: More than 5 species of Rhizophoraceae grow in Australasia with particularly high biodiversity on the island of New Guinea and northern Australia.

Australia has about 11,500 km^2 of mangroves, primarily on the northern and eastern coasts of the continent, with occurrences as far south as Millers Landing in Wilsons Promontory, Victoria (38°542 S) and Barker Inlet in Adelaide, South Australia.

New Zealand: New Zealand also has mangrove forests extending to around 38°S (similar to Australia's southernmost mangrove incidence): the furthest geographical extent on the west coast is Raglan Harbour (37°482 S); on the east coast, Ohiwa Harbour (near Opotiki) is the furthest south mangroves are found (38°002 S).

Pacific Islands

Twenty-five species of mangrove are found on various Pacific islands, with extensive mangals on some islands. Mangals on Guam, Palau, Kosrae and Yap have been badly affected by development.

Mangroves are not native to Hawaii, but the red mangrove, *Rhizophora mangle*, and Oriental mangrove, *Bruguiera sexangula*, have been introduced and are now naturalized. Both species are considered invasive species and classified as pests by the University of Hawaii Botany Department.

Exploitation and Conservation

Approximately 35% of mangrove area was lost during the last several decades of the 20th century (in countries for which sufficient data exist), which encompass about half of the area of mangroves. The United Nations Environment Program & Hamilton (2013), estimate that shrimp farming causes approximately a quarter of the destruction of mangrove forests. Likewise, the 2010 update of the World Mangrove Atlas indicated a fifth of the world's mangrove ecosystems have been lost since 1980.

Figure: *Mangroves in West Bali National Park, Indonesia.*

Grassroots efforts to save mangroves from development are becoming more popular as their benefits become more widely known. In the Bahamas, for example, active efforts to save mangroves are occurring on the islands of Bimini and Great Guana Cay. In Trinidad and Tobago as well, efforts are underway to protect a mangrove threatened by the construction of a steelmill and a port. In Thailand, community management has been effective in restoring damaged mangroves. Within northern Ecuador mangrove regrowth is reported in almost all estuaries and stems primarily from local actors responding to earlier periods of deforestation in the Esmeraldas region.

Mangroves have been reported to be able to help buffer against tsunami, cyclones, and other storms. One village in Tamil Nadu was protected from tsunami destruction - the villagers in Naluvedapathy planted 80,244 saplings to get into the Guinness Book of World Records. This created a kilometre-wide belt of trees of various varieties. When the tsunami struck, much of the land around the village was flooded, but the village itself suffered minimal damage.

Reforestation

Figure: *Mangroves in Bohol, Philippines.*

In some areas, mangrove reforestation and mangrove restoration is also underway. Red mangroves are the most common choice for cultivation, used particularly in marine aquariums in a sump to reduce nitrates and other nutrients in the water. Mangroves also appear in home aquariums, and as ornamental plants, such as in Japan.

In Senegal, Haïdar El Ali has started the fr project, which (amongst others) focuses on reforesting several areas with mangroves.

The Manzanar Mangrove Initiative is an ongoing experiment in Arkiko, Eritrea, part of the Manzanar Project founded by Gordon H.

Sato, establishing new mangrove plantations on the coastal mudflats. Initial plantings failed, but observation of the areas where mangroves did survive by themselves led to the conclusion that nutrients in water flow from inland were important to the health of the mangroves. Trials with the Eritrean Ministry of Fisheries followed, and a planting system was designed to introducing the nitrogen, phosphorus, and iron missing from seawater.

The propagules are planted inside a reused galvanized steel can with the bottom knocked out; a small piece of iron and a pierced plastic bag with fertilizer containing nitrogen and phosphorus are buried with the propagule. As of 2007, after six years of planting, 700,000 mangroves are growing; providing stock feed for sheep and habitat for oysters, crabs, other bivalves, and fish.

National Studies

In terms of local and national studies of mangrove loss, the case of Belize's mangroves is illustrative in its contrast to the global picture. A recent, satellite-based study - funded by the World Wildlife Fund and conducted by the Water Center for the Humid Tropics of Latin America and the Caribbean (CATHALAC) – indicates Belize's mangrove cover declined by a mere 2% over a 30-year period. The study was born out of the need to verify the popular conception that mangrove clearing in Belize was rampant.

Instead, the assessment showed, between 1980 and 2010, under 4,000 acres (16 km^2) of mangroves had been cleared, although clearing of mangroves near Belize's main coastal settlements (e.g. Belize City and San Pedro) was relatively high. The rate of loss of Belize's mangroves - at 0.07% per year between 1980 and 2010 - was much lower than Belize's overall rate of forest clearing (0.6% per year in the same period). These findings can also be interpreted to indicate Belize's mangrove regulations (under the nation's) have largely been effective. Nevertheless, the need to protect Belize's mangroves is imperative, as a 2009 study by the World Resources Institute (WRI) indicates the ecosystems contribute US$174–249 million per year to Belize's national economy.

Blue Carbon

Blue carbon is the carbon captured by the world's oceans and coastal ecosystems. The carbon captured by living organisms in oceans is stored in the form of biomass and sediments from mangroves, salt marshes and seagrasses.

The rates of blue carbon sequestration and storage capacities in ecosystems are comparable to (and often higher than) those in carbon-rich terrestrial ecosystems such as tropical rainforests or peatlands. Unlike most terrestrial systems, which reach soil carbon equilibrium within decades, deposition of carbon dioxide in coastal ecosystem sediment can continue over millennia. However, when these coastal ecosystems are degraded or destroyed they can become carbon dioxide sources due to the oxidization of biomass and organic soil.

Because coastal ecosystems do contain substantial amounts of carbon, and because this carbon is in danger of being released, they are important in mitigating climate change. However, the rate of loss of mangroves, sea grasses and salt marshes (driven mostly by human activities) is estimated to be among the highest of any ecosystem on the planet, prompting international interest in managing them more effectively for their carbon benefits.

These ecosystems are highly valuable not only for their contribution to climate change mitigation on a global scale, but also for the many valuable services they provide locally.

Various processes are known to enhance the ocean's ability to store carbon. Sperm whales increase the levels of primary production and carbon export to the deep ocean by depositing iron rich faeces into surface waters of the Southern Ocean. The iron rich faeces causes phytoplankton to grow and take up more carbon from the atmosphere. When the phytoplankton dies, it sinks to the deep ocean and takes the atmospheric carbon with it. By reducing the abundance of sperm whales in the Southern Ocean, whaling has resulted in an extra 2 million tonnes of carbon remaining in the atmosphere each year.

Sundarban

The Sundarbans is a natural region in Bengal. It is the largest single block of tidal halophytic mangrove forest in the world. The Sundarbans covers approximately 10,000 square kilometres (3,900 sq mi) of which 60 percent is in Bangladesh with the remainder in India. The Sundarbans is a UNESCO World Heritage Site.

The Sundarbans National Park is a National Park, Tiger Reserve, and a Biosphere Reserve located in the Sundarbans delta in the Indian state of West Bengal. Sundarbans South, East and West are three protected forests in Bangladesh. This region is densely covered by mangrove forests, and is one of the largest reserves for the Bengal tiger.

Geography

The Sundarban forest lies in the vast delta on the Bay of Bengal formed by the super confluence of the Padma, Brahmaputra and Meghna rivers across southern Bangladesh. The seasonally flooded Sundarbans freshwater swamp forests lie inland from the mangrove forests on the coastal fringe. The forest covers 10,000 square kilometres (3,900 sq mi) of which about 6,000 square kilometres (2,300 sq mi) are in Bangladesh. It became inscribed as a UNESCO world heritage site in 1997. The Indian part of Sundarbans is estimated to be about 4,110 square kilometres (1,590 sq mi), of which about 1,700 square kilometres (660 sq mi) is occupied by waterbodies in the forms of river, canals and creeks of width varying from a few meters to several kilometres.

The Sundarbans is intersected by a complex network of tidal waterways, mudflats and small islands of salt-tolerant mangrove forests. The interconnected network of waterways makes almost every corner of the forest accessible by boat. The area is known for the eponymous Royal Bengal tiger (*Panthera tigris tigris*), as well as numerous fauna including species of birds, spotted deer, crocodiles and snakes. The fertile soils of the delta have been subject to intensive human use for centuries, and the ecoregion has been mostly converted to intensive agriculture, with few enclaves of forest remaining. The remaining forests, taken together with the Sundarbans mangroves, are important habitat for the endangered tiger. Additionally, the Sundarbans serves a crucial function as a protective barrier for the millions of inhabitants in and around Khulna and Mongla against the floods that result from the cyclones. The Sundarbans has also been enlisted among the finalists in the New7Wonders of Nature.

Physiography

The mangrove-dominated Ganges Delta – the Sundarbans – is a complex ecosystem comprising one of the three largest single tracts of mangrove forests of the world. Situated mostly in Bangladesh, a small portion of it lies in India. The Indian part of the forest is estimated to be about 19 percent, while the Bangladeshi part is 81 percent. To the south the forest meets the Bay of Bengal; to the east it is bordered by the Baleswar River and to the north there is a sharp interface with intensively cultivated land. The natural drainage in the upstream areas, other than the main river channels, is everywhere impeded by extensive embankments and polders. The Sundarbans was originally measured (about 200 years ago) to be of about 16,700 square kilometres (6,400 sq mi). Now it has dwindled into about 1/3 of the original size.

The total land area today is 4,143 square kilometres (1,600 sq mi), including exposed sandbars with a total area of 42 square kilometres (16 sq mi); the remaining water area of 1,874 square kilometres (724 sq mi) encompasses rivers, small streams and canals. Rivers in the Sundarbans are meeting places of salt water and freshwater. Thus, it is a region of transition between the freshwater of the rivers originating from the Ganges and the saline water of the Bay of Bengal.

The Sundarbans along the Bay of Bengal has evolved over the millennia through natural deposition of upstream sediments accompanied by intertidal segregation. The physiography is dominated by deltaic formations that include innumerable drainage lines associated with surface and subaqueous levees, splays and tidal flats. There are also marginal marshes above mean tide level, tidal sandbars and islands with their networks of tidal channels, subaqueous distal bars and proto-delta clays and silt sediments. The Sundarbans' floor varies from 0.9 to 2.11 metres (3.0 to 6.9 ft) above sea level.

Biotic factors here play a significant role in physical coastal evolution, and for wildlife a variety of habitats have developed which include beaches, estuaries, permanent and semi-permanent swamps, tidal flats, tidal creeks, coastal dunes, back dunes and levees. The mangrove vegetation itself assists in the formation of new landmass and the intertidal vegetation plays a significant role in swamp morphology. The activities of mangrove fauna in the intertidal mudflats develop micromorphological features that trap and hold sediments to create a substratum for mangrove seeds. The morphology and evolution of the eolian dunes is controlled by an abundance of xerophytic and halophytic plants. Creepers, grasses and sedges stabilise sand dunes and uncompacted sediments. The Sunderbans mudflats (Banerjee, 1998) are found at the estuary and on the deltaic islands where low velocity of river and tidal current occurs. The flats are exposed in low tides and submerged in high tides, thus being changed morphologically even in one tidal cycle. The interior parts of the mudflats are magnificent home of luxuriant mangroves.

Ecoregions

Sundarbans features two ecoregions — "Sundarbans freshwater swamp forests" (IM0162) and "Sundarbans mangroves" (IM1406).

Sundarbans Freshwater Swamp Forests

The Sundarbans freshwater swamp forests are a tropical moist broadleaf forest ecoregion of Bangladesh. It represents the brackish swamp forests that lie behind the *Sundarbans Mangroves*, where the salinity is more pronounced. The freshwater ecoregion is an area where the water is only slightly brackish and becomes quite fresh during the rainy season, when the freshwater plumes from the Ganges and the Brahmaputra rivers push the intruding salt water out and bring a deposit of silt. It covers an area of 14,600 square kilometres (5,600 sq mi) of the vast Ganges-Brahmaputra Delta, extending from the northern part of Khulna District and finishing at the mouth of the Bay of Bengal with scattered portions extending into India's West Bengal state. The Sundarbans freshwater swamp forests lie between the upland Lower Gangetic plains moist deciduous forests and the brackish-water Sundarbans mangroves bordering the Bay of Bengal.

A victim of large-scale clearing and settlement to support one of the densest human populations in Asia, this ecoregion is under a great threat of extinction. Hundreds of years of habitation and exploitation have exacted a heavy toll on this ecoregion's habitat and biodiversity. There are two protected areas – Narendrapur (110 km^2) and Ata Danga Baor (20 km^2) that cover a mere 130 km^2 of the ecoregion. Habitat loss in this ecoregion is so extensive, and the remaining habitat is so fragmented, that it is difficult to ascertain the composition of the original vegetation of this ecoregion. According to Champion and Seth (1968), the freshwater swamp forests are characterised by *Heritiera minor*, *Xylocarpus molluccensis*, *Bruguiera conjugata*, *Sonneratia apetala*, *Avicennia officinalis*, and *Sonneratia caseolaris*, with *Pandanus tectorius*, *Hibiscus tiliaceus*, and *Nipa fruticans* along the fringing banks.

Sundarbans Mangroves

The Sundarbans Mangroves ecoregion on the coast forms the seaward fringe of the delta and is the world's largest mangrove ecosystem, with 20,400 square kilometres (7,900 sq mi) of area covered. The dominant mangrove species *Heritiera fomes* is locally known as sundri or sundari. Mangrove forests are not home to a great variety of plants. They have a thick canopy, and the undergrowth is mostly seedlings of the mangrove trees. Besides the sundari, other species that make up the forest include *Avicennia spp.*, *Xylocarpus mekongensis*, *Xylocarpus granatum*, *Sonneratia apetala*, *Bruguiera gymnorrhiza*, *Cereops decandra*, *Aegiceras corniculatum*, *Rhizophora mucronata*, and *Nypa fruticans* palms.

Twenty-six of the fifty broad mangrove types found in the world grow well in the Sundarbans. The commonly identifiable vegetation that grow in the dense mangrove forests at the Sundarbans are salt water mixed forest, mangrove scrub, brackish water mixed forest, littoral forest, wet forest and wet alluvial grass forests. The Bangladesh mangrove vegetation of the Sundarbans differs greatly from other non-deltaic coastal mangrove forests and upland forests associations. Unlike the former, the Rhizophoraceae are of minor importance.

Ecological Succession

Ecological succession is generally defined as the successive occupation of a site by different plant communities. In an accreting mudflats the outer community along the sequence represents the pioneer community which is gradually replaced by the next community representing the seral stages and finally by a climax community typical of the climatic zone. Troup suggested that succession began in the newly accreted land created by fresh deposits of eroded soil. The pioneer vegetation on these newly accreted sites is *Sonneratia*, followed by *Avicennia* and *Nypa*. As the ground is elevated as a result of soil deposition, other trees make their appearance. The most prevalent, though one of the late species to appear, is *Excoecaria*. As the level of land rises through accretion and the land is only occasionally flooded by tides, *Heritiera fomes* begins to appear.

Flora

A total 245 genera and 334 plant species were recorded by David Prain in 1903. While most of the mangroves in other parts of the world are characterised by members of the Rhizophoraceae, Avicenneaceae or Combretaceae, the mangroves of Bangladesh are dominated by the Malvaceae and Euphorbiaceae. Dominant flora includes:

The Sundarbans flora is characterised by the abundance of sundari (*Heritiera fomes*), gewa (*Excoecaria agallocha*), goran (*Ceriops decandra*) and keora (*Sonneratia apetala*) all of which occur prominently throughout the area. The characteristic tree of the forest is the *sundari* (*Heritiera littoralis*), from which the name of the forest had probably been derived. It yields a hard wood, used for building houses and making boats, furniture and other things. New forest accretions is often conspicuously dominated by keora (*Sonneratia apetala*) and tidal forests. It is an indicator species for newly accreted mudbanks and is an important species for wildlife, especially spotted deer (*Axis axis*). There is abundance of dhundul or passur (*Xylocarpus granatum*) and kankra (*Bruguiera gymnorrhiza*) though distribution is discontinuous. Among palms, *Poresia coaractata*, *Myriostachya wightiana* and golpata (*Nypa fruticans*), and among grasses spear grass (*Imperata cylindrica*) and khagra (*Phragmites karka*) are well distributed.

The varieties of the forests that exist in Sundarbans include mangrove scrub, littoral forest, saltwater mixed forest, brackish water mixed forest and swamp forest. Besides the forest, there are extensive areas of brackish water and freshwater marshes, intertidal mudflats, sandflats, sand dunes with typical dune vegetation, open grassland on sandy soils and raised areas supporting a variety of terrestrial shrubs and trees. Since Prain's report there have been considerable changes in the status of various mangrove species and taxonomic revision of the man-grove flora. However, very little exploration of the botanical nature of the Sundarbans has been made to keep up with these changes. Differences in vegetation have been explained in terms of freshwater and low salinity influences in the Northeast and variations in drainage and siltation. The Sundarbans has been classified as a moist tropical forest demonstrating a whole mosaic of seres, comprising primary colonisation on new accretions to more mature beach forests. Historically vegetation types have been recognised in broad correlation with varying degrees of water salinity, freshwater flushing and physiography.

Fauna

The Sundarbans provides a unique ecosystem and a rich wildlife habitat. According to the 2011 tiger census, the Sundarbans have about 270 tigers. Although previous rough estimates had suggested much higher figures close to 300, the 2011 census provided the first ever scientific estimate of tigers from the area Tiger attacks are frequent in the Sundarbans. Between 100 and 250 people are killed per year.

There is much more wildlife here than just the endangered Royal Bengal tiger (*Panthera tigris tigris*). Most importantly, mangroves are a transition from the marine to freshwater and terrestrial systems, and provide critical habitat for numerous species of small fish, crabs, shrimps and other crustaceans that adapt to feed and shelter, and reproduce among the tangled mass of roots, known as pneumatophores, which grow upward from the anaerobic mud to get the supply of oxygen. Fishing cats, macaques, wild boars, common grey mongooses, foxes, jungle cats, flying foxes, pangolins, and spotted deer are also found in abundance in the Sundarbans.

A 1991 study has revealed that the Bangladeshi part of the Sundarbans supports diverse biological resources including at least 150 species of commercially important fish, 270 species of birds, 42 species of mammals, 35 reptiles and 8 amphibian species. This represents a significant proportion of the species present in Bangladesh (i.e. about 30% of the reptiles, 37% the birds and 34% of the mammals) and includes a large number of species which are now extinct elsewhere in the country. Two amphibians, 14 reptiles, 25 aves and five mammals are endangered. The Sundarbans is an important wintering area for migrant water birds and is an area suitable for watching and studying avifauna.

The management of wildlife is restricted to, firstly, the protection of fauna from poaching, and, secondly, designation of some areas as wildlife sanctuaries where no extraction of forest produce is allowed and where the wildlife face few disturbances. Although the fauna of Bangladesh have diminished in recent times and the Sundarbans has not been spared from this decline, the mangrove forest retains several good wildlife habitats and their associated fauna. Of these, the tiger and dolphin are target species for planning wildlife management and tourism development. There are high profile and vulnerable mammals living in two contrasting environments, and their statuses and management are strong indicators of the general condition and management of wildlife. Some of the species are protected by legislation, notably by the Bangladesh Wildlife (Preservation) Order, 1973 (P.O. 23 of 1973).

Predators

The fertile soils of the delta have been subject to intensive human use for centuries, and the ecoregion has been mostly converted to intensive agriculture, with few enclaves of forest remaining. The remaining forests, together with the Sundarbans mangroves, are important habitats for the endangered Bengal tiger (*Panthera tigris*). The forest also contains leopard (*Panthera pardus fusca*) and several other smaller predators such as the jungle cats (*Felis chaus*), fishing cats (*Prionailurus viverrinus*), and leopard cats (*Prionailurus bengalensis*).

Several predators dwell in the labyrinth of channels, branches and roots that poke up into the air. This is the only mangrove ecoregion that harbours the Indo-Pacific region's largest predator, the Bengal tiger. Unlike in other habitats, tigers live here and swim among the mangrove islands, where they hunt scarce prey such as the chital deer (*axis axis*), Indian muntjacs (*Muntiacus muntjak*), wild boars (*Sus scrofa*), and even rhesus macaque (*Macaca mulatta*). It is estimated that there are now 500 Bengal tigers and about 30,000 spotted deer in the area. The tigers regularly attack and kill humans who venture into the forest, human deaths ranging from 30–100 per year.

Some of the reptiles are predators too, including two species of crocodiles, the saltwater crocodile (*Crocodylus porosus*) and mugger crocodile *(Crocodylus palustris)*, as well as the gharial (*Gavialis gangeticus*) and the water monitor lizards (*Varanus salvator*), all of which hunt on both land and water. Sharks and the Gangetic dolphins (*Platanista gangetica*) roam the waterways.

Avifauna

The forest is also rich in bird life, with 170 species including the endemic brown-winged kingfishers *(Pelargopsis amauroptera)* and the globally threatened lesser adjutants *(Leptoptilos javanicus)* and masked finfoots *(Heliopais personata)* and birds of prey such as the ospreys *(Pandion haliaetus)*, white-bellied sea eagles *(Haliaeetus leucogaster)* and grey-headed fish eagles *(Ichthyophaga ichthyaetus)*. The Sundarbans was designated a Ramsar site on 21 May 1992. Some of the more popular birds found in this region are open billed storks, black-headed ibis, water hens, coots, pheasant-tailed jacanas, pariah kites, Brahminy kites, marsh harriers, swamp partridges, red junglefowls, spotted doves, common mynahs, jungle crows, jungle babblers, cotton teals, herring gulls, Caspian terns, gray herons, Brahminy ducks, spot-billed pelicans, great egrets, night herons, common snipes, wood sandpipers, green pigeons, rose-ringed parakeets, paradise flycatchers, cormorants, white-bellied sea eagles, seagulls, common kingfishers, peregrine falcons, woodpeckers, whimbrels, black-tailed godwits, little stints, eastern knots, curlews, golden plovers, pintails, white-eyed pochards and lesser whistling ducks.

Aqua Fauna

Some of the fish and amphibians found in the Sundarbans are sawfish, butter fish, electric ray, common carp, silver carp, barb, river eels, starfish, king crab, fiddler crab, hermit crab, prawn, shrimps, Gangetic dolphins, skipping frogs, common toads and tree frogs. One particularly interesting fish is the mudskipper, a gobioid that climbs out of the water into mudflats and even climbs trees.

Reptiles

The Sundarbans National Park houses an excellent number of reptiles as well. Some of the common ones are olive ridley turtles, sea snakes, dog faced water snakes, green turtles, estuarine crocodiles, chameleons, king cobras, salvator lizards, hard shelled batgun terrapins, Russels vipers, mouse gekkos, monitor lizards, curviers, hawks bill turtles, pythons, common kraits, green vine snake, checkered keelbacks and rat snakes. The river terrapin (*Batagur baska*), Indian flap-shelled turtles (*Lissemys punctata*), peacock soft-shelled turtles (*Trionyx hurum*), yellow monitors (*Varanus flavescens*), water monitors (*Varanus salvator*), and Indian pythons (*Python molurus*) are some of the resident species.

Endangered and Extinct Species

Forest inventories reveal a decline in standing volume of the two main commercial mangrove species – sundari (*Heritiera* spp.) and gewa (*Excoecaria agallocha*) — by 40% and 45% respectively between 1959 and 1983. Despite a total ban on all killing or capture of wildlife other than fish and some invertebrates, it appears that there is a consistent pattern of depleted biodiversity or loss of species (notably at least six mammals and one important reptile) in the 20th century, and that the "ecological quality of the original mangrove forest is declining".

The endangered species that live within the Sundarbans and extinct species that used to be include the Royal Bengal tigers, estuarine crocodile, northern river terrapins (*Batagur baska*), olive ridley sea turtles, Gangetic dolphin, ground turtles, hawksbill sea turtles and king crabs (horse shoe). Some species such as hog deer (*Axis porcinus*), water buffalos (*Bubalus bubalis*), barasingha or swamp deer (*Cervus duvauceli*), Javan rhinoceros (*Rhinoceros sondaicus*), single horned rhinoceros (*Rhinoceros unicornis*) and the mugger crocodiles or marsh crocodiles (*Crocodylus palustris*) started to become extinct in the Sundarbans towards the middle of the 20th century, due to extensive poaching and man hunting by the British. There are several other threatened mammal species, such as the capped langurs (*Semnopithecus pileatus*), smooth-coated otters (*Lutrogale perspicillata*), Oriental small-clawed otters (*Aonyx cinerea*), and great Bengal civets (*Viverra zibetha*).

Climate Change Impact

The physical development processes along the coast are influenced by a multitude of factors, comprising wave motions, micro and macro-tidal cycles and long shore currents typical to the coastal tract. The shore currents vary greatly along with the monsoon. These are also affected by cyclonic action. Erosion and accretion through these forces maintains varying levels, as yet not properly measured, of physiographic change whilst the mangrove vegetation itself provides a remarkable stability to the entire system. During each monsoon season almost all the Bengal Delta is submerged, much of it for half a year. The sediment of the lower delta plain is primarily advected inland by monsoonal coastal setup and cyclonic events. One of the greatest challenges people living on the Ganges Delta may face in coming years is the threat of rising sea levels caused mostly by subsidence in the region and partly by climate change.

In many of the Bangladesh's mangrove wetlands, freshwater reaching the mangroves was considerably reduced from the 1970s due to diversion of freshwater in the upstream area by neighbouring India through the use of the Farakka Barrage bordering Rajshahi, Bangladesh. Also, the Bengal Basin is slowly tilting towards the east due to neo-tectonic movement, forcing greater freshwater input to the Bangladesh Sundarbans. As a result, the salinity of the Bangladesh Sundarbans is much lower than that of the Indian side. A 1990 study noted that there "is no evidence that environmental degradation in the Himalayas or a 'greenhouse' induced rise in sea level have aggravated floods in Bangladesh"; however, a 2007 report by UNESCO, "Case Studies on Climate Change and World Heritage" has stated that an anthropogenic 45-centimetre (18 in) rise in sea level (likely by the end of the 21st century, according to the Intergovernmental Panel on Climate Change), combined with other forms of anthropogenic stress on the Sundarbans, could lead to the destruction of 75 percent of the Sundarbans mangroves. Already, Lohachara Island and New Moore Island/South Talpatti Island have disappeared under the sea, and Ghoramara Island is half submerged.

Economy

The Sundarbans plays an important role in the economy of the southwestern region of Bangladesh as well as in the national economy. It is the single largest source of forest produce in the country. The forest provides raw materials for wood based industries. In addition to traditional forest produce like timber, fuelwood, pulpwood etc., large scale harvest of non-wood forest products such as thatching materials, honey, bees-wax, fish, crustacean and mollusc resources of the forest takes place regularly. The vegetated tidal lands of the Sundarbans also function as an essential habitat, produces nutrients and purifies water. The forest also traps nutrient and sediment, acts as a storm barrier, shore stabiliser and energy storage unit. Last but not the least, the Sunderbans provides a wonderful aesthetic attraction for local and foreign tourists.

The forest also has immense protective and productive functions. Constituting 51% of the total reserved forest estate of Bangladesh, it contributes about 41% of total forest revenue and accounts for about 45% of all timber and fuel wood output of the country. A number of industries (e.g. newsprint mill, match factory, hardboard, boat building, furniture making) are based on the raw materials obtained from the Sundarbans ecosystem. Various non-timber forest products and plantations help generate considerable employment and income generation opportunities for at least half a million poor coastal population. It also provides natural protection to life and properties of the coastal population in the cyclone prone Bangladesh.

Habitation

The Sundarbans has a population of over 4 million but much of it is mostly free of permanent human habitation. Despite human habitations and a century of economic exploitation of the forest well into the late forties, the Sundarbans retained a forest closure of about 70% according to the Overseas Development Administration (ODA) of the United Kingdom in 1979.

Administration

The Sundarbans area is one of the most densely populated in the world, and the population is increasing. As a result, half of this ecoregion's mangrove forests have been cut down to supply fuelwood and other natural resources. Despite the intense and large-scale exploitation, this still is one of the largest contiguous areas of mangroves in the world. Another threat comes from deforestation and water diversion from the rivers inland, which causes far more silt to be brought to the estuary, clogging up the waterways.

The Directorate of Forest is responsible for the administration and management of Sundarban National Park in West Bengal. The Principal Chief Conservator of Forests (PCCF), Wildlife & Bio-Diversity & ex-officio Chief Wildlife Warden, West Bengal is the senior most executive officer looking over the administration of the park. The Chief Conservator of Forests (South) & Director, Sundarban Biosphere Reserve is the administrative head of the park at the local level and is assisted by a Deputy Field Director and an Assistant Field Director. The park area is divided into two ranges, overseen by range forest officers. Each range is further sub-divided into beats. The park also has floating watch stations and camps to protect the property from poachers.

The park receives financial aid from the State Government as well as the Ministry of Environment and Forests under various Plan and Non-Plan Budgets. Additional funding is received under the Project Tiger from the Central Government. In 2001, a grant of US$20,000 was received as a preparatory assistance for promotion between India and Bangladesh from the World Heritage Fund.

A new Khulna Forest Circle was created in Bangladesh back in 1993 to preserve the forest, and Chief Conservators of Forests have been posted since. The direct administrative head of the Division is the Divisional Forest Officer, based at Khulna, who has a number of professional, subprofessional and support staff and logistic supports for the implementation of necessary management and administrative activities. The basic unit of management is the compartment. There are 55 compartments in four Forest Ranges and these are clearly demarcated mainly by natural features such as rivers, canals and creeks.

Protection

The Bangladesh part of the forest lies under two forest divisions, and four administrative ranges viz Chandpai (Khulna District), Sarankhola (Khulna), and Burigoalini (Satkhira District) and has sixteen forest stations. It is further divided into fifty-five compartments and nine blocks. There are three wildlife sanctuaries established in 1977 under the Bangladesh Wildlife (Preservation) Order, 1973 (P.O. 23 of 1973). The West Bengal part of the forest lies under the district of South & North 24 Parganas.

Protected areas cover 15% of the Sundarbans mangroves including Sundarbans National Park and Sajnakhali Wildlife Sanctuary, in West Bengal, Sundarbans East, Sundarbans South and Sundarbans West Wildlife Sanctuaries in Bangladesh.

Sundarban National Park

The Sundarban National Park is a National Park, Tiger Reserve, and a Biosphere Reserve in West Bengal, India. It is part of the Sundarbans on the Ganges Delta, and adjacent to the Sundarbans Reserve Forest in Bangladesh. The delta is densely covered by mangrove forests, and is one of the largest reserves for the Bengal tiger. It is also home to a variety of bird, reptile and invertebrate species, including the salt-water crocodile. The present Sundarbans National Park was declared as the core area of Sundarbans Tiger Reserve in 1973 and a wildlife sanctuary in 1977. On 4 May 1984 it was declared a National Park.

Sundarbans West Wildlife Sanctuary

Sundarbans West Wildlife Sanctuary is a UNESCO World Heritage Site. The region supports several mangroves, including: sparse stands of Gewa (*Excoecaria agallocha*) and dense stands of Goran (*Ceriops tagal*), with discontinuous patches of Hantal palm (*Phoenix paludosa*) on drier ground, river banks and levees. The fauna of the sanctuary is very diverse with some 40 species of mammals, 260 species of birds and 35 species of reptiles. The greatest of these being the Bengal Tiger of which an estimated 350 remain in the Bangladesh Sundarbans. Other large mammals are wild boar, Chital horin (spotted deer), Indian otter and macaque monkey. Five species of marine turtles frequent the coastal zone and two endangered reptiles are present – the estuarine crocodile and the Indian python.

Sundarbans East Wildlife Sanctuary

Sundarbans East Wildlife Sanctuary extends over an area of 31,227 hectares (77,160 acres). Sundari trees (*Heritiera fomes*) dominate the flora, interspersed with Gewa (*Excoecaria agallocha*) and Passur (*Xylocarpus mekongensis*) with Kankra (*Bruguiera gymnorrhiza*) occurring in areas subject to more frequent flooding. There is an understory of Shingra (*Cynometra ramiflora*) where, soils are drier and Amur (*Aglaia cucullata*) in wetter areas and Goran (*Ceriops decandra*) in more saline places. Nypa palm (*Nypa fruticans*) is widespread along drainage lines.

Sundarbans South Wildlife Sanctuary

Sundarbans South Wildlife Sanctuary extends over an area of 36,970 hectares (91,400 acres). There is evidently the greatest seasonal variation in salinity levels and possibly represents an area of relatively longer duration of moderate salinity where Gewa (*Excoecaria agallocha*) is the dominant woody species. It is often mixed with Sundri, which is able to displace in circumstances such as artificially opened canopies where Sundri does not regenerate as effectively. It is also frequently associated with a dense understory of Goran (*Ceriops tagal*) and sometimes Passur.

Sajnakhali Wildlife Sanctuary

Sajnakhali Wildlife Sanctuary is a 362-square-kilometre (140 sq mi) area in the northern part of the Sundarbans delta in South 24 Parganas district, West Bengal, India. The area is mainly mangrove scrub, forest and swamp. It was set up as a sanctuary in 1976. It is home to a rich population of different species of wildlife, such as water fowl, heron, pelican, spotted deer, Rhesus Macaques, wild boar, tigers, Water Monitor Lizards, Fishing Cats, otters, Olive ridley turtle, crocodiles, Batagur Terrapins, and migratory birds.

Mineral Resources of Coastal Tracts

"Minerals" form the backbone of the industries and consequently provide necessary synergy by their application in various manufacturing sectors. Therefore, there is an ever lasting symbiotic relationship between the minerals and their industrial products. The GDP of a country is measured in terms of its industrial output. Therefore, optimal usage of the mineral resources is one of the key factors among others in determining the GDP of any country. This is true not only of "developing countries" like India but equally applicable to the industrially developed countries such as USA, Germany, France, England, China, etc. If India is to progress and march ahead to catch up with other developed countries we should endeavour to focus on harnessing its mineral resources into industrial purposes. By planned utilisation of the available mineral resources, the locally available resources can not only be put into various industrial uses but the surplus mineral resources can be exported as well thereby earning valuable foreign exchange to the country. The fact that mining activity provides gainful employment to several lakhs of remote rural people where mineral occurrences are located can not be lost sight of.

Placer, beach minerals sands in the coastal areas, unlike the aforesaid techniques adopted in respect of mineral deposits mentioned above the occurrence of beach mineral sands / placer minerals in the coastal area is unique in nature with its occurrence getting confined to a very shallow depth ranging from the surface 2 to 3 meters depth only. This is the only known replenishable natural resource in the whole of the minerals scenario. The mining technique in winning these placer minerals is very simple using shovel, spade, etc. No blasting is required unlike the other open cast or other under ground mining. The mining technique adopted is almost analogous to sand mining from the river bed. This is in complete contrast to the highly skilful and scientific method of mining adopted in open cast as well as under ground mining. In these categories of mining blasting of varying intensity has to be adopted.

Heavy Mineral Sands Ore Deposits

Heavy mineral sands are a class of ore deposit which is an important source of zirconium, titanium, thorium, tungsten, rare earth elements, the industrial minerals diamond, sapphire, garnet, and occasionally precious metals or gemstones.

Heavy mineral sands are placer deposits formed most usually in beach environments by concentration due to the specific gravity of the mineral grains. It is equally likely that some concentrations of heavy minerals (aside from the usual gold placers) exist within streambeds, but most are of a low grade and are relatively small.

The grade of a typical heavy mineral sand ore deposit is usually low. Within the 21st century, the lowest cut-off grades of heavy minerals, as a total heavy mineral (THM) concentrate from the bulk sand, in most ore deposits of this type is around 1% heavy minerals, although several are higher grade.

Of this total heavy mineral concentrate (THM), the components are typically

- Zircon, from 1% of THM to upwards of 50% of THM,
- Ilmenite, generally of 10% to 60% of THM
- Rutile, from 5% to 25% of THM
- Leucoxene, from 1% to 10% of THM
- Trash minerals, typically quartz, magnetite, garnet, chromite and kyanite, which usually account for the remaining bulk of the THM content
- Slimes, typically minerals as above and heavy clay minerals, too fine to be economically extracted.

Generally, as zircon is the most valuable component and a critical ore component, high-zircon sands are the most valuable. Thereafter, rutile, leucoxene and then ilmenite in terms of value given to the ore. As a generality, typically the valuable components of the THM concentrate rarely exceed 30%.

Being ancient stranded dune systems, the tonnage of most deposits is in excess of several tens of millions of tonnes to several hundred million tonnes. For example, the medium-sized Coburn mineral sands deposit, Western Australia, is 230 million tonnes at 1.1% heavy minerals, and is 13 km long.

Sources

The source of heavy mineral sands is in a hardrock source within the erosional areas of a river which carries its load of sediment into the ocean, where the sediments are caught up in littoral drift or longshore drift. Rocks are occasionally eroded directly by wave action shed detritus, which is caught up in longshore drift and washed up onto beaches where the lighter minerals are winnowed.

The source rocks which provide the heavy mineral sands determine the composition of the economic minerals. The source of zircon, monazite, rutile, sometimes tungsten, and some ilmenite is usually granite. The source of ilmenite, garnet, sapphire and diamond is ultramafic and mafic rocks, such as kimberlite or basalt. Garnet is also sourced commonly from metamorphic rocks, such as amphibolite schists. Precious metals are sourced from ore deposits hosted within metamorphic rocks.

Transport

The accumulation of a heavy mineral deposit requires a source of sediment containing heavy minerals onto a beach system in a volume which exceeds the rate of removal from the trap site. For this reason not all beaches which are supplied by sands containing heavy minerals will form economic concentrations of the minerals. This factor can be qualitatively or quantitatively measured through the ZTR index.

Trap

The heavy minerals within the source sediments attain an economic concentration by accumulation within low-energy environments within streams and most usually on beaches. In beach placer deposits the lowest energy zone on the beach is the swash zone, where turbulent surf washes up on the beach face and loses energy. In this zone, heavier grains accumulate because they are denser than the quartz grains they occur with and become stranded. It is for this reason that beach placer deposits are often referred to as "strand-line deposits".

The size and position of a heavy mineral deposit is a function of the wave energy reaching the beach, the mean grainsize of the beach sediments, and the current height of the ocean.

Anecdotal reports of certain beach placers which are forming in modern times suggest that the greatest enrichment of the sands tends to occur in storm events which are energetic enough to remove most of the beaches sediment load—a process favoring the lighter minerals. The resultant 'clinker' sands which were left behind were mined during low tide following major storm events, suggesting that most beach placer deposits are formed during such cycles.

In most cases, fossilised dune systems are exploited for heavy mineral sands because they are from the ocean and because they are often remnants of previous intraglacial highstands.

Tectonic activity, which results in coastlines rising from the ocean, may also cause a beach system to become stranded above the high-water mark and lock in the heavy mineral sands. Similarly, a beach system which is drowned by the subsidence of a coastline may be preserved, often for millions of years, until it is either covered by sedimentation or rises from the ocean.

Specific trap sites for heavy mineral sand placer deposits are in beaches on the leeward side of headlands, as this forms a low-energy zone which traps sediments carried along by the longshore drift. Also, sand bars developed at the mouths of rivers which feed the placer deposits are rich trap sites where the winnowing action of the waves are most efficient, because heavy minerals, if they are going to be too heavy to be moved, will deposit at an isthmus in preference to drifting too far down the beach.

Diamond Sands

The coast of Namibia is host to economic diamantiferous beach sands, which are exploited by building sea walls and isolating stretches of coastline. The beaches are so isolated that they are sometimes processed in their entirety, down to the bedrock, in search of diamonds. Such deposits have been sought around the world, with sporadic reports of high-value stones but no instances of economic quantities of sediment.

Environmental Concerns

The mining of beach sands and of fossilized beach placers is often controversial because the operation requires the strip mining of large areas. Often this land is in ecologically sensitive surroundings and contains fragile ecosystems built up on poor sandy soils.

The mining process is ideally modelled on the extraction operations underway in Australia, where the strip mining is followed by rehabilitation of the mined areas including intensive re-vegetation with ecologically similar species, re-contouring of the land to its original shape, including dunes, and management of groundwater resources. Modern mining practices tend to favor dry mining rather than dredging operations, due to the advent of electrostatic mineral separation processes.

In practice, not all mining of sub-Saharan African deposits is carried out in such an environmentally responsible manner, although some South African mines do practice dune rehabilitation. The mining of the coast of South America, in particular Chile and Ecuador, is carried out in an environmentally responsible manner.

Examples of environmentally sensitive and politically sensitive mineral sands mining operations which have gained public attention and galvanised environmental activism responses to mining proposals include the Tuart-Ludlow mineral sands mine, Western Australia, and the culmination of conservationist efforts to preserve Rainbow Beach and Fraser Island, Queensland, Australia. These latter campaigns successfully lobbied government and saw Fraser Island and Rainbow Beach protected by the High Court of Australia; however, the Tuart-Ludlow campaign failed to protect the Tuart forests in coastal Western Australia.

Coastal and Marine Fisheries Resources of Kerala

The coast of Kerala constitutes approximately 10 percent of India's total coastline which provide opportunities in traditional fishing in inshore waters from ages. The coastline of Kerala, considered as the Exclusive Economic Zone (EEZ) which extends up to 200 nautical miles far beyond the continental shelf offers lucrative fishery.

The Southwest coastal (SWC) has certain unique features that influence the fishery fluctuations of the important commercial species to a great extent. The area is subjected to two monsoons viz. the south-west monsoon (Edvapathi) and the north-east monsoon (Thulavarsham). The southwest monsoon coincides with the period of upwelling and phytoplankton bloom, which results in a large number of fish and crustaceans in the area. The maximum utilization of this continental shelf is only in 50m depth ie. around 22 km from the coastline.

South – west monsoon coupled with northwesterly winds and the oceanic currents cause upwelling along the coast brings the nutrient rich deep waters to the surface, which flourishes the primary production and followed by a good fishery. Kerala coast have major fisheries of the elasmobranches, cat fish, shrimps, cuttle fish, sardines, anchovilla, saurida & saurus, perches, sciaenids, mackerels, ribbon fish, seer fish, tunis, prawn, anchovies, soles, sharks, rays, etc. Kerala ranks first in marine fish production of India forming nearly 25% (average 5.75 lakh tonnes) of the total annual production. Currently the annual export of marine products from the state yields to the nation a foreign exchange of Rs. 1,100 crores.

There has been spectacular growth in the marine fisheries sector of the state due to fisheries friendly government policies, well developed harvest and post harvest infrastructure and increased demand for seafood both in the domestic and export markets. Kerala has been in the forefront in absorbing innovative and new technologies in fishing practices and adoption of these technologies has led to marine fisheries to take a complex structure.

The growing demand resulted in fishing intensification, extension of fishing grounds, increase in overall length, fish holding capacity of the trawlers, fishing effort in terms of fishing hours through multi-day fishing by the mechanized sector and enhanced fishing operations by the motorized sector especially the ring seiners. The intensification of mechanized fishing, perceived as a threat to the sustenance of traditional fisheries sector and resource sustainability, culminated in the regulation of mechanized fishing activities through the Kerala

Marine Fisheries Regulation Act 1980, and thereby enforced a ban on bottom trawling during the monsoon season since 1988.

And also the introduction of ban on trawling coincided with introduction of the highly efficient mass harvesting gear, namely the ringseine, by the traditional motorised sector, which resulted in a quantum leap in the total pelagic fish production. In the context of globalization of trade, growing demand for seafood, enormous pressure on the resources and increased awareness of ecosystem-based management and eco-labeling, there is an urgent need for developing a management regime based on the principles of responsible fishing for ensuring livelihood security, resource sustainability, economic efficiency and ecosystem integrity.

Fishing industry which has grown substantially in the last four decades with the continuous intensification of fishing effort, contributes significantly to economic, social and nutritional well being of the people.

Major technological advancements in fishing crafts and gears during the past four decades of Kerala are

- developments in craft technology and mechanization of propulsion, gear and catch handling
- introduction of synthetic gear materials
- developments in acoustic fish detection and satellite-based remote sensing techniques
- advances in electronic navigation and position fixing equipment
- awareness of the need for responsible fishing to ensure sustainability of the resources, protection of the biodiversity and environmental safety and energy efficiency.

Major technological changes that have taken place in the capture fisheries of Kerala are the

- Introduction and popularization of synthetic fishing gear materials.
- Introduction of trawling in mid1950s
- Improvement in efficiency and diversification of trawls, purse seines, gillnets and lines, for mechanized sector.
- Expansion of fishing grounds for harvesting deep sea fishing for deeps sea prawns, lobsters and cephalopods.
- Continuous improvement in size, endurance, installed engine power, winch capacities, fish-hold, freshwater and fuel capacities of mechanized trawlers to enable multi-day fishing, since mid 1980s.

- Improvement in size, endurance, installed engine power, fish-hold, freshwater and fuel capacities of mechanized gillnetters/liners to enable multi-day, distant water fishing.
- Adoption of modern technologies such as echo sounder and GPS in a wider scale over the last decade.
- Motorization of traditional fishing crafts and expansion in fishing grounds
- Improvement of traditional fishing units, in terms of craft modernization, gear materials, gear efficiency and dimensions.
- Introduction of ring seines in commercial fishing in 1986
- Displacement of traditional boat seines by ring seines.
- Introduction of mini-trawling in mid-1987 and its subsequent proliferation, targeting near-shore shrimp and fish resources.
- Introduction of ring seines with inboard engines and purse line haulers in 1999 and continuous increase in numbers.

Coastal Fishery Resources of India with Special Reference

Coastal fish, also called offshore fish or neritic fish, inhabit the sea between the shoreline and the edge of the continental shelf. Since the continental shelf is usually less than 200 metres deep, it follows that pelagic coastal fish are generally epipelagic fish, inhabiting the sunlit epipelagic zone. Coastal fish can be contrasted with ocean fish, which inhabit the oceans beyond the continental shelves.

Coastal fish are the most abundant in the world. They can be found in tidal pools, fjords and estuaries, near sandy shores and rocky coastlines, around coral reefs and on or above the continental shelf. Coastal fish include forage fish and the predator fish that feed on them. Forage fish thrive in inshore waters where high productivity results from upwelling and shoreline run off of nutrients. Some are partial residents that spawn in streams, estuaries and bays, but most complete their life cycles in the zone.

Coastal fish are found in the waters above the continental shelves that extend from the continental shorelines, and around the coral reefs that surround volcanic islands. The total world shoreline extends for 356,000 km (221,000 mi) and the continental shelves occupy a total area of 24.286 million km^2 (9 376 million sq mi). This is about 4.8% of the world's total area of 510.072 million km^2.

Nearshore Fish

Nearshore fish, sometimes called littoral fish, live close to the shore. They are associated with the intertidal zone, or with estuaries, lagoons, coral reefs, kelp forests, seagrass meadows, or rocky or sandy bottoms, usually in shallow waters less than about 10 m deep.

Intertidal Fish

Intertidal fish are fish that move in and out with the tide in the intertidal zone of the seashore, or are found in rock pools or under rocks.

The intertidal zone of rocky shores can contain indentations which trap pools of salty water, called rock pools. Living in these habitats are communities of hardy plant and animal species specially adapted for coping with the volatile environment around them. The plants and animals interact with each other and with the rock pool to form miniature ecosystems, easily accessible to students and a source of fascination for young children. Plants such as seaweeds, cnidarians such as sea anemones, arthropods like barnacles, and molluscs such as the common limpet and the common periwinkle can be permanent residents of rock pools. But most rock pool animals, such as crabs, shrimp and fish are just temporary residents, occupying a rock pool only until the next tide takes them to a new location.

Some rock pool fish which are temporary residents include the long-spined sea scorpion, the pipefish worm, the rock goby and the common lumpsucker. However some other rock pool fish are territorial in nature, and will stay with the same pool for extended periods. Examples are the common blenny and its near relative the butterfish.

- The common blenny, also known as the shanny, is found in northern temperate waters. They hide under rocks and in crannies in rock pools when the tide is out. They feed on green seaweed and invertebrates such as barnacles. They can crawl on dry land, using their paired fins. About 16 cm long, they have smooth skin, without scales, and are covered with soft slime. The slime prevents them drying if they are stranded on a shore between tides. So long as their skin stays moist, they can breathe out of water. They are sometimes called "sea frogs" because they bask in the sun on weeds outside the water, and like frogs, jump to safety when disturbed. They can change their colour to match their surroundings. The female lays eggs in crevices or under stones and the male guards them until they hatch. In the winter, when storms can be severe, they move out of their rock pools into the shallows. The common blenny is bold with strong teeth, and will bite humans if it feels threatened.

- The rock goby is a small fish, about 12 cm long, found in northern temperate waters. It is coloured black with white blotches, and hides under stones and amongst seaweed. It is a temporary resident of rock pools when the tide is out. The female rock goby lays eggs on the underside of rocks and shells and then leaves them. The male guards the eggs until they hatch. First-year rock gobies often visit rock pools in winter when the older fish have left.
- The long-spined sea scorpion, a small stout fish which grows about 29 cm long, is another temporary resident of rocky pools. They have large black eyes, a large mouth, and four long spines—two on each side on the gill cover—that stick out when the fish is removed from the water. They also have an organ like a finger on each side of their mouths which helps them catch prey. Because of their broad heads, they are also called "bullheads". They have a variety of effective camouflaged colours ranging from shades of browns with cream blotches, to orange and red with white blotches. They can also change their body colour to match their surroundings. They are found around the coasts of Northern Europe in shallow rocky waters hiding amongst seaweed. They are also found in rock pools and sometimes in waters 30 m deep. Long-spined sea scorpions lay eggs amongst seaweed or attached to rock crevices. The young hatch after two or three weeks, and go through several development stages before maturing into adults.
- Lumpsuckers are found in temperate northern waters. They live on the seafloor, and are temporary residents of rocky pools in late winter and early spring when they spawn. The body of the lumpsucker is scaleless and covered with small lumps. They have a large sucking disc on their underside which they use to cling to surfaces. They are normally a blue to slate-grey colour, and are effectively camouflaged to look like stones. They are portly, nearly spherical, poor swimmers, reaching lengths up to 50 cm. After the female lumpsucker lays eggs, the male takes over, clamping itself to a rock where it guards the eggs. When they hatch, lumpsuckers look like tiny tadpoles. They remain in shallow water and rock pools, hiding amongst seaweed and rocks, until they grow up.

Estuarine Fish

Estuaries are partly enclosed coastal bodies of water with one or more rivers or streams flowing into them, and with a free connection to the open sea. These brackish water habitats form a transition zone between river environments and ocean environments, and ecological successions can form along the way. Estuaries are subject to both marine influences, such as tides, waves, and the influx of saline water; and riverine influences, such as flows of fresh water and sediment. The inflow of both seawater and freshwater provide high levels of nutrients in both the water column and sediment, making estuaries productive natural habitats.

Fishes that spend time in estuaries (or river mouths) need to be euryhaline (tolerant to a range of salinities). Estuaries provide an unstable environment for fish, where the salinity changes and the waters are often muddy and turbulent. In warmer climates, estuaries have mangroves around their edges. At times there may be only a few different fish species present in an estuary, but seasonal migrants, including eels, salmonids, and some forage fish such as herrings and sprats increase the diversity in the estuary.

River estuaries form important staging points during the migration of anadromous and catadromus fish species, such as salmon and eels, giving them time to form social groups and to adjust to the changes in salinity. Salmon are anadromous, meaning they live in the sea but ascend rivers to spawn; eels are catadromous, living in rivers and streams, but returning to the sea to breed. Besides the species that migrate through estuaries, there are many other fish that use them as "nursery grounds" for spawning or as places young fish can feed and grow before moving elsewhere. For example, herring and plaice are two commercially important species that use the Thames Estuary for this purpose.

Mangrove swamps are associated brackish water habitats. Many, though not all, mangrove swamps fringe estuaries and lagoons where the salinity changes with each tide. Among the most specialised residents of mangrove forests are mudskippers, fish that forage for food on land, and archer fish, perch-like fish that "spit" at insects and other small animals living in the trees, knocking them into the water where they can be eaten. Like estuaries, mangrove swamps are important breeding grounds for many fish, with species such as snappers, halfbeaks, and tarpon spawning or maturing among them.

Coral Reef Fish

In tropical waters, coral reef fish live amongst or in close relation to coral reefs. Coral reefs form complex ecosystems with tremendous biodiversity. Coral reef fish can be particularly colourful and interesting to watch. Hundreds of species can exist in a small area of a healthy reef, many of them hidden or well camouflaged. Reef fish have developed many ingenious specialisations adapted to survival on the reefs. Coral reefs occupy less than one percent of the surface area of the world oceans, yet they provide a home for 25 percent of all marine fish species.

Coral reefs often depend on other habitats in the surrounding area for the supply of nutrients, such as seagrass meadows and mangrove forests. Seagrass and mangroves supply dead plants and animals which are rich in nitrogen and also serve to feed fish and animals from the reef by supplying wood and vegetation. Reefs in turn protect mangroves and seagrass from waves and produce sediment for the mangroves and seagrass to root in.

Anthias are members of the family Serranidae (basses, basslets, groupers) and make up the subfamily Anthiinae. They are widespread in tropical waters. They have been called the "quintessential reef fish", and make up a sizeable portion of the colourful fishes seen swarming in coral reef photography. Anthias are mostly small, peaceful, beautiful and popular as ornamental fish. They are mainly zooplankton feeders. Anthias shoal and school in large numbers, operating more intimate "harems" within the schools. These harems contain a dominant and colourful male, between 2 and 12 females — who operate a hierarchy among themselves — and one or two "subdominant" males, often less brightly coloured and non-territorial. Within the swarm of females, territorial males perform acrobatic U-swim displays and vigorously defend an area of the reef and its associated harem. Anthias are protogynous hermaphrodites. All anthias are born female; if a dominant male perishes, the largest female of the group will often change into a male to take its place. This may lead to squabbling between the next largest male and the transforming female, whose hormones are now surging with testosterone. This can turn quite vicious in the limited confines of captivity.

Butterflyfish are group of about 120 species belonging to the family Chaetodontidaeof Perchiformes. They include bannerfish and coralfish. They are widespread on coral reefs. Butterflyfish are mostly between 12 and 22 centimetres (4.7 and 8.7 in) in length. The largest species, the lined butterflyfish and saddle butterflyfish, grow to 30 centimetres (12 in). Many species are brightly coloured and strikingly patterned, though other species are dull in colour. Many have eyespots on their flanks and dark bands across their eyes, not unlike the patterns seen on butterfly wings. Their deep, laterally narrow bodies are easily noticed through the profusion of reef life. The conspicuous colouration of butterflyfish may be intended for interspecies communication. Butterflyfish have uninterrupted dorsal fins with tail fins that may be rounded or truncated, but are never forked. Generally diurnal and frequenting waters of less than 18 metres (59 ft) (though some species descend to 180 metres (590 ft)), butterflyfish stick to particular home ranges. The corallivores are especially territorial, forming mated pairs and staking claim to a specific coral head. Contrastingly, the zooplankton feeders form large conspecific groups. By night butterflyfish hide in reef crevices and exhibit markedly different colouration. Their colouration also makes butterflyfish popular aquarium fish. However, most species feed on coral polyps and sea anemones, which can result in problems for the hobby aquarists.

Clownfish, anemonefish and damselfish are among about 360 species classified in the family Pomacentridae. Most Pomacentrids are associated with coral reefs in the Indo-West Pacific, with a few species occurring in temperate waters. Some species are native to freshwater or brackish estuarine environments. Most live in shallow water, from 2 to 15 meters (6 to 50 feet), although some species are found below 100 m (330 feet). Most species are specialists, living in specific parts of the reef, such as sandy lagoons, steep reef slopes, or areas exposed to strong wave action. In general, the coral is used as shelter, and many species can only survive in its presence. The bottom-dwelling species are territorial, occupying and defending a portion of the reef, often centred around an area of shelter. By keeping away other species of fish, some pomacentrids encourage the growth of thick mats of algae within their territories, leading to the common name farmerfish. Different species display a wide range of colours, although some are relatively drab. Pomacentrids are omnivorous or herbivorous, feeding off algae, plankton, and small bottom-dwelling crustaceans. A small number eat coral.

Goatfishes are a family Mullidae of about 55 species of perciform fishes, associated worldwide with tropical reefs. They are typically about 20 cm long, though the dash-and-dot goatfish, grows to 55 cm. Goatfish are tireless benthic feeders, possess a pair of long chemosensory barbels ("whiskers") protruding from their chins resembling a goat's beard. They use these to rifle through the sediments in search of a meal.

Like goats, they seek anything edible; worms, crustaceans, molluscs and other small invertebrates are staples. Many species of goatfish are conspicuously coloured and have the ability to change their colouration depending on their current activity. By day, many form large inactive (non-feeding) schools: these aggregates may contain both conspecifics and heterospecifics.

For example, the yellowfin goatfish school with blue-striped snappers. When they do that, the yellowfins changes its colouration to match that of the snapper. By night the schools disperse and individual goatfish head their separate ways to loot the sands. The diurnal goldsaddle goatfish changes from a lemon-yellow to a pale cream when feeding. Other nocturnal feeders will shadow the active goatfish, waiting patiently for overlooked morsels. Goatfish stay within the shallows, going no deeper than about 110 metres. Most species do not tolerate brackish water, so they do not enter estuaries or the mouths of rivers.

Other Nearshore Fish

Other nearshore or shallow water fish live near the shore in depths of less than 10 metres. They occupy the areas over sandy or rocky bottoms, and can be associated with seagrass meadows and kelp forests. They can be divided into demersal fish and pelagic fish. Demersal fish live on or near the sea floor, while pelagic fish live in the water column away the sea floor.

Examples of such shallow water demersal fish, found in both tropical and temperate waters around the world, are triplefins, seahorses, wrasse and flounder. As demersal fish, all these fish spend most of their time on or near the sea floor.

- Flatfish are superbly adapted groundfish, found on muddy and sandy sea floors. In many species both eyes lie on one side of the head, one or the other migrating through and around the head during development. Some species face their "left" side upward, some face their "right" side upward, and others face either side upward. Some flatfish can camouflage themselves on the ocean floor.
- Wrasse are a large family of mainly small fish, usually less than 20 centimetres (7.9 in) long. Most wrasse are loners that prefer habitats such as coral reefs and rocky shores. They live close to the substrate, eating small invertebrates and almost anything else that lurks on the bottom. Many are brightly coloured. They have thick lips and use their sharp teeth to pick small creatures off the rocks. Many smaller wrasses follow the feeding trails of larger fish, picking up invertebrates disturbed by their passing.
- Triplefins are a family of fish. They are usually found around coral reefs and rocks, usually in shallow, clear sunlit waters such as lagoons and seaward reefs. Triplefins have three dorsal fins (hence the name). They are small fish, usually less than six cm long. Brightly coloured, often for reasons of camouflage, they are nervous and retreat to rock crevices at any perceived threat.
- Seahorses are a genus of fish. They prefer sheltered harbours, estuaries and other shallow coastal waters, where they hunt tiny crustaceans. They bob around in sheltered areas such as coral reefs, mangrove stands and seagrass meadows and estuaries. They are camouflaged with murky patterns that blend into kelp and sea grass backgrounds. During social moments or in unusual surroundings, seahorses can turn on bright colours.

Examples of shallow water pelagic fish, found in both tropical and temperate waters around the world, are grey mullet, sprats and garfish. As pelagic fish, all these fish spend most of their time living in the water column away the sea floor.

- The grey mullet are medium size fish, typically about 50 cm (20 in) long. They are often caught with set nets.
- The garfish is a long, slender fish, looking like a spear, which feeds on seagrass fragments, shrimps and crab larvae. In turn it is preyed on by larger fish and, since it is often near the surface, cormorants and gannets.

Plankton Feeding Pelagic Fish

At the base of food chains are the primary producers. In the ocean these primary producers are mainly a type of plankton, microscopic phytoplankton which drift in the water column. Phytoplankton need sunlight for photosynthesis to power carbon fixation, so they are mainly located in sunlit surface waters. Phytoplankton also need and rapidly use nutrients in the water column. The phytoplankton are eaten by zooplankton, which in turn are eaten by predatory zooplankton. Filter feeders then eat the plankton and larger predatory fish eat the filter feeders.

Most filter feeding pelagic fish found in coastal waters are small, silvery forage fish. Forage fish include fishes of the family Clupeidae (herrings, shad, sardines and pilchards, hilsa, menhaden and sprats), as well as anchovies, capelin and halfbeaks. They use schooling strategies to avoid predators, and different schools of forage fish often associate with each other in open coastal waters. Forage fish feed near the base of the food chain on plankton and fry (recently hatched fish), often by filter feeding. In turn, they are preyed on by larger predatory fish, seabirds and marine mammals.

Worldwide, there are five major coastal currents associated with upwelling areas: the Canary Current (off Northwest Africa), the Benguela Current (off southern Africa), the California Current (off California and Oregon), the Humboldt Current (off Peru and Chile), and the Somali Current (off Western India). All of these currents support major fisheries. Many forage fish are important commercial species, and the schools can be targeted by spotter planes. The fish are caught by purse seiners—fishing boats that use nets to enclose the fish—and can be overfished.

Predatory Pelagic Fish

Predatory pelagic fishes found on continental shelves worldwide in both tropical and temperate waters include porgies, barracuda, amberjacks and cutlassfishes. They tend to be larger fish, and are carnivorous, feeding on the smaller, silvery forage fish that eat plankton. Some species also feed on crabs and other invertebrates, foraged from the sea floor.

- Mackerel scom*
- Porgies sometimes called sea breams, are any of about 100 species belonging to the family Sparidae. Porgies usually have high backs and a single dorsal fin, like snapper or grunt fishes (grunts are named for the sound they make grinding their teeth).

They are bottom feeding pelagic fishes, with small mouths equipped with strong teeth adapted for handling small fishes and invertebrates with hard shells. Most do not exceed a size of about 30 cm (12 in), but some may grow to four times that length. They often school, and will migrate between reefs. Larger fish enter estuaries and harbours.

- Barracuda have long slender bodies typically about 50 cm (20 in) long. They have a wicked set of teeth and are ferocious predators. They feed on crustaceans, cephalopods and small fish like anchovy and pilchard. Barracouta often hunt in schools near the bottom or midwater, and sometimes even near the surface at night.
- Cutlassfishes are a group of about 40 species belonging to the family Trichiuridae. They are ocean fish which regularly stray into coastal waters around the world. Fish of this family are long, slender, and generally steely blue or silver in colour, giving rise to their name. They have reduced or absent pelvic and caudal fins, giving them an eel-like appearance, and large fang-like teeth.
- Jacks, amberjacks, pompanos, horse mackerel, scads, leatherjackets and trevally are fish of the family *Carangidae*. Found in most coastal waters, they are fast predatory fishes that hunt in the waters above reefs and in the open sea; some dig in the sea floor for invertebrates (some can also filter feed, such as the white trevally). The largest fish in the family, the giant trevally, grows up to 1.7 m in length; most fish in the family reach a maximum length of 25–100 cm. The family contains many important commercial and game fish, notably the Pacific jack mackerel and the other jack mackerels in the genus *Trachurus*. The type species of this genus is the Atlantic horse mackerel. Jack mackerels are an important inshore commercial species.
  - o Amberjacks are a group of nine species belonging to the genus *Seriola* within the family Carangidae. Mainly open water fish, they can follow small forage fish into estuaries and enclosed waters, where they will also hunt for crustaceans. Amberjacks are fast swimming and aggressive predators that often hunt in schools around offshore reefs. The yellowtail amberjack can reach 1.8 metres in length and weigh 60 kilograms.

Demersal Fish

Fish that live on or in close association with the sea floor are called demersal fish. This section discusses the coastal demersal fish that live on the continental shelf, but are living further from the coast and in deeper water than the nearshore fish discussed above.

Demersal fish are white fish. Unlike oily fish, white fish contain oils only in their liver, rather than in the gut, and can therefore be gutted as soon as they are caught, on board the ship. White fish has dry and white flesh. They can be divided into benthopelagic fish (mostly "round" fish) which live *near* the sea bed, such as cod, and benthic fish (flatfish) such as plaice which live *on* the sea bed. Benthic fish tend to be "flat", so they can lie on the bottom.

- Cod-like fishes are a number of round benthopelagic species belonging to the order Gadiformes, such as Atlantic and Pacific cod, morid cod, haddock and pollock, including the highly commercial Alaska pollock. Cod-like fishes are often found in large schools over sandy or muddy bottoms. They have a barbel (fleshy filament) on their lower jaw which they use to detect prey buried in the sand or mud. Some migrate to warm water in winter to spawn.
- John dory are fishes of the genus *Zeus*. They have a widespread distribution and are typically found near the seabed in depths from 5 meters (15 ft). The John dory grows to a maximum length of 65 cm (2 ft). Although it is a benthopelagic fish, its body is flat and it can hardly be seen from the front because it is so thin. It is a poor swimmer with long spines on the dorsal fin. It has a large dark eyespot on the side of its body which is used to confuse prey, which are scooped up in its big mouth. Large eyes at the front of the head provide it with bifocal vision and depth perception, which are important for predators. The John Dory usually gets its food by stalking it then shooting out a tube in its mouth to capture its prey. It eats forage fish, and occasionally squid and cuttlefish. In turn, they are preyed on by sharks, like the dusky shark, and other large bony fish. They are normally solitary.
- Turbot and brill are benthic flatfish, resembling flounder and sole, but found in deeper offshore waters on the continental shelf. They are brownish-green, with dark blotches on the turbot and mottling on the brill. They are fished by coastal trawlers.

Mail-cheeked fishes belong to a group of about 30 species in the order Scorpaeniformes. Mail-cheeked fishes are named after a plate of bone that runs across each cheek. They are widespread in all the oceans of the world. Mail-cheeked fishes are carnivorous, mostly feeding on crustaceans, such as crabs and shrimp, and on smaller fish. Most species live on the sea bottom in relatively shallow waters, although species are known from mid and deep water, from the mid-water, and even from fresh water. They typically have spiny heads, and rounded pectoral and caudal fins. Most species are less than 30 centimetres (12 in) in length, but the full size range of the order varies from the velvetfishes, which can be just 2 centimetres (0.79 in) long as adults, to the Lingcod, which can reach 150 centimetres (4.9 ft) in length.

Red gurnard are mail-cheeked fish. They use their large pectoral fins to rest on the bottom and to detect food.

Stargazers are about 50 species of fishes, belonging to the family Uranoscopidae, and found worldwide in shallow waters. Stargazers are venomous; they have two large poison spines situated behind the opercle and above the pectoral fins. They can also deliver electric shocks. They are ambush predators with eyes on top of their heads (thus the name). Stargazers also have a large upward-facing mouth in a large head. They bury themselves in sand with only their eyes showing, and leap upwards to ambush fish and invertebrates overhead. Some species have a worm-shaped lure growing out of the floor of the mouth, which they wiggle to attract prey's attention. Lengths range from 18 cm up to 90 cm, for the giant stargazer *Kathetostoma giganteum*. Stargazers are a delicacy in some cultures. The venom is destroyed when it is cooked, and stargazers are sold in some fish markets with their electric organ removed. They have been called "the meanest things in creation" and the "worst pet on earth".

Sandperches are a family, Pinguipedidae, containing 63 species of fishes in the order Perciformes. They are benthic carnivoures, feeding on small fish and invertebrates. Examples are the redbanded weever, yellow weaver and blue cod. They are often caught in pots like crayfish.

Medusa fishes are a family Centrolophidae of 31 species of perciform fishes. They are found in temperate and tropical waters throughout the world, usually feeding on fish, crustaceans and small squid near rough sea floors on continental shelf and slope. Examples are barrelfish, southern driftfish, imperial blackfish, the Japanese and pelagic butterfish, the New Zealand and Tasmanian ruffe, and the common, silver and white warehouse. The young of some species associate with jellyfish, which provides them with protection from predators and opportunities to scavenge the remains of the jellyfish's meals. The young of other species associate with large masses of floating kelp.

Grouper are fish belonging to a number of genera in the subfamily Epinephelinae of the family Serranidae, in the order Perciformes. Species of grouper include the black, comet, gag, giant, Goliath, Nassau, saddletail, tiger, Warsaw, white and yellowfin grouper. Typical lengths are 80–120 centimetres. They inhabit depths from reefs near the surface down to over 400 metres. They feed on just about any moving animal they encounter. Groper are important inshore commercial fish, usually caught with gill nets (in earlier times longlines were used).

Wreckfish are a family Polyprionidae of perciform fishes, found on the floor of the continental shelf and slope where they inhabit caves and shipwrecks (thus their common name). The Atlantic wreckfish is at depths between 40 and 600 m (130 to 2,000 ft). They are largely a solitary fish, though juveniles school below floating objects. Their diet includes large ocean cephalopods, crustaceans, and other bottom-dwelling fishes.

An estuary is formed where a river meets the sea. Some estuaries are permanently open to the sea, while others seldom open. The salinity (saltiness) of estuaries can vary greatly and depends on the mixing of the sea with freshwater from the river. Estuaries are constantly changing and are influenced by the tide, waves and rainfall. The big changes in temperature, salinity and oxygen concentration make estuaries a very harsh environment to live in.

Despite this, many types of plants and animals have adapted to live in estuaries. Plants such as algae, eelgrass, saltmarsh plants, mangroves and reeds thrive in this environment. Estuaries are important nursery areas for many fish and almost 100 species of fish are partially or completely dependent on estuaries at some stage of their lives. Common estuarine fish species include the spotted grunter, dusky kob, Cape stumpnose and several species of mullet.

Many birds depend on estuaries for food. These include fish-eaters such as the fish eagle, herons and cormorants, invertebrate-eaters such as flamingos and weed-eaters such as red-knobbed coots. Estuarine environments are threatened by the construction of dams, agriculture, property development and pollution.

Status of Marine Fisheries in India

Fisheries is an important sector in India—it provides employment to millions of people and contributes to food security of the country.

With a coastline of over 8,000 km, an Exclusive Economic Zone (EEZ) of over 2 million sq km, and with extensive freshwater resources, fisheries play a vital role. Presently, fisheries and aquaculture contribute 1.07 per cent to the national GDP, and 5.30 per cent to agriculture and allied activities, while the average annual value of output during the Tenth Five Year Plan (2002-2007) was Rs31,682.50 crores.

Marine Fisheries contributes to food security and provides direct employment to over 1.5 mn fisher people besides others indirectly dependent on the sector.

According to the CMFRI Census 2010, there are 3,288 marine fishing villages and 1,511 marine fish landing centres in 9 maritime states and 2 union territories. The total marine fisherfolk population was about 4 million comprising in 864,550 families. Nearly 61% of the fishermen families were under BPL category. The average family size was 4.63 and the overall sex ratio was 928 females per 1000 males. The Indian coastline can be delineated into 22 zones, based on the ecosystem structure and functions. The Indian boat type ranges from the traditional catamarans, masula boats, plank-built boats, dug out canoes, machwas, dhonis to the present day motorized fibre-glass boats, mechanized trawlers and gillnetters. In the marine fisheries sector, there were 194,490 crafts in the fishery out of which 37% were mechanized, 37% were motorized and 26% were non-motorized. Out of a total of 167,957 crafts fully owned by fisherfolk 53% were non-motorized, 24% were motorized and 23% were mechanized. Among the mechanized crafts fully owned by fishermen, 29% were trawlers, 43% were gillnetters and 19% were dolnetters.

Chapter 6

Longterm Morphological Changes of Shoreline and Estuaries

Shoreline changes induced by erosion and accretion are natural processes that take place over a range of time scales. They may occur in response to smaller-scale (short-term) events, such as storms, regular wave action, tides and winds, or in response to large-scale (long-term) events such as glaciation or orogenic cycles that may significantly alter sea levels (rise/fall) and tectonic activities that cause coastal land subsidence or emergence. Hence, most coastlines are naturally dynamic, and cycles of erosion are often an important feature of their ecological character. Wind, waves and currents are natural forces that easily move the unconsolidated sand and soils in the coastal area, resulting in rapid changes in the position of the shoreline.

Excluding the impact of human activity, these processes are simply natural evolutionary phenonema. Human activities along the coast (land reclamation, port development, shrimp farming), within river catchments and watersheds (river damming and diversion) and offshore (dredging, sand mining) in combination with these natural forces often exacerbate coastal erosion in many places and jeopardize opportunities for coasts to fulfill their socio-economic and ecological roles in the long term at a reasonable societal cost.

Development within coastal areas has increased interest in erosion problems; it has led to major efforts to manage coastal erosion problems and to restore coastal capacity to accommodate short-and long-term changes induced by human activities, extreme events and sea level rise. The erosion problem becomes worse whenever the countermeasures (i.e. hard or soft structural options) applied are inappropriate, improperly designed, built, or maintained and if the effects on adjacent shores are not carefully evaluated. Often erosion is addressed locally at specific places or at regional or jurisdictional boundaries instead of at system boundaries that reflect natural processes. This anomaly is mostly attributable to insufficient knowledge of coastal processes and the protective function of coastal systems.

The costs of installing hard structures for coastal protection are very high; strong negative public reaction to rock emplacements along the coast often aggravate the problem. This has led to uncertainty among managers and local government authorities on how to treat shoreline erosion. It has become an issue for serious debate for politicians, coastal managers, land- and property owners, lawyers, bankers, insurers and fisher folk, especially in areas of intensive use and rapidly rising coastal land value. Many of these stakeholders are resorting to planned retreat where houses or hotels are simply removed and the coast is left to erode. However, planned retreat can be expensive, unnecessary and sometimes impossible, especially in highly modified environments.

Increased interest in soft structures for coastal protection (including increased forest cover) and a combination of hard and soft structures is predominating and is consonant with advanced knowledge on coastal processes and natural protective functions. There is evidence that coastal forests and trees provide some coastal protection and that the clearing of coastal forests and trees has increased the vulnerability of coasts to erosion — such as in Viet Nam, Malaysia, Indonesia, Sri Lanka, India China and Thailand. This paper will elaborate on and discuss the causes of coastal erosion induced by human activities; erosion management options; and the role of coastal forests and trees in protecting coastal areas against coastal erosion, as well as their socio-economic and environmental considerations.

Coastal erosion and accretion are natural processes; however, they have become anomalous and widespread in the coastal zone of Asia and other countries in the Indian Ocean owing to combinations of various natural forces, population growth and unmanaged economic development along the coast, within river catchments and offshore. This type of erosion has been reported in China, Japan, India, Indonesia, Viet Nam, Sri Lanka, Thailand, Bangladesh and Malaysia.

The Extent of the Coastal Erosion Problem in Asia

Bilan reported that the erosion rate in the northern part of Jiangsu Province in China is serious and as high as 85 metres/year; in Hangzhou

Bay the rate is 40 metres/year, while in Tianjin it is 16–56 metres/year. Erosion persists even where preventive measures such as sea dykes are constructed. Beach scour has been found along coasts with sea-dyke protection.

This erosion is attributable to many factors such as river damming and diversion, that leads to less sediment supply to the coast, and the clearing of mangrove forests, which makes coastal areas more susceptible to the hazard. Juxtaposing these phenomena, the intensification of typhoons and storm surges during the 42-year period between 1949 and 1990 has meant that storm surges with increasing tidal levels exceeding one and two metres have occurred 260 and 48 times respectively, thus exacerbating the erosion problem. Most of the sediment taken offshore by the storm waves has been returned in minimal quantities to the coast during normal conditions owing to the frequent storm intensity.

According to Othman, nearly 30 percent of the Malaysian coastline is undergoing erosion. Many of these areas are coastal mudflats, fringed by mangroves. Behind the mangroves there are usually agricultural fields protected from tidal inundation by bunds (dykes). Locally, mangroves are known to reduce wave energy as waves travel through them; thus, the Department of Irrigation and Drainage has ruled that at least 200 metres of mangrove belts must be kept between the bunds and the sea to protect the bunds from eroding. However, the mangroves themselves are susceptible to erosion when the soil under their root systems is undermined by wave action that mostly occurs during periods of lower water level or low tide. The value of intact mangrove swamps for storm protection and flood control alone in Malaysia is approximately US$300 000/kilometre.

In Viet Nam, most of the coastline in the south that is located in a wide and flat alluvial fan and bordered by tidal rivers fringed by wide mangrove swamps, has been eroded continuously at a rate of approximately 50 metres/year since the early twentieth century. This massive erosion — mostly due to wave and current action — and vanishing mangrove vegetation is attributable to the long-term impacts of human activities since the late nineteenth century and also human-induced change within watersheds (dam construction that has reduced the sediment supply to the shore). Erosion still occurs in the central coastal zone of Viet Nam and preventive measures such as sea dykes, revetments, and tree plantations have been implemented in many coastal areas; however, in the southern coastal zone, mangrove plantations have mitigated wave action and prevented further erosion.

The rapid erosion of the coast of Sagar Island in West Bengal, India, is caused by several processes that act in concert; these are natural processes that occur frequently (cyclones, waves and tides that can reach six metres in height) and anthropogenic activities such as human settlement and aquaculture that remove mangroves and other coastal vegetation. The erosion rate from 1996 to 1999 was calculated to be 5.47 square kilometres/year. The areas that are severely affected by erosion are the northeastern, southwestern and southeastern faces of the island. Malini and Rao (2004) reported coastal erosion and habitat loss along the Godavari Delta front owing to the combination of the dam construction across the Godavari and its tributaries that diminishes sediment supply to the coast and continued coastal land subsidence.

Sri Lanka's experience with coastal erosion dates back to 1920. It has become more serious because mangroves are being eradicated by encroachment (human settlement), fuelwood cutting and the clearing of coastal areas for intensive shrimp culture. Mangrove forest cover was estimated to be approximately 12 000 hectares in 1986; this dwindled to 8 687 hectares in 1993 and was estimated to be only 6 000 hectares in 2000. Approximately US$30 million has already been spent on breakwaters and other construction to combat coastal erosion on southern and western coasts; however, coastal erosion still persists in some coastal areas.

In Indonesia, coastal erosion started in the northern coast of Java Island in the 1970s when most of the mangrove forest had been converted to shrimp ponds and other aquaculture activities, and the area was also subjected to unmanaged coastal development, diversion of upland freshwater and river damming. Coastal erosion is prevalent throughout many provinces such as Lampung, Northeast Sumatra, Kalimantan, West Sumatra (Padang), Nusa Tenggara, Papua, South Sulawesi and Bali. US$79.667 million was provided by the Indonesian Government to combat coastal erosion from 1996 to 2004, but only for Bali Island in order to protect this valuable coastal tourism base. A combination of hard structures and engineering approaches (breakwaters/jetties/revetments) of different shapes that fused functional design and aesthetic values, and soft structures and engineering approaches (beach nourishment) was used. They succeeded in stopping coastal erosion on Sanur, Nusa Dua and Tanjong Benoa beaches, but were neither cost effective nor efficient, because during low tide all of the coastal area was exposed up to 300 metres offshore; thus, these huge structures were revealed and became eyesores.

In Thailand, intensification of coastal erosion came to notice during the past decade. Overall, the net erosion is approximately 1.3 to 1.7 metres/year along the southern Thailand coastline. Total area losses amount to 0.91 square kilometres/year for the Gulf coast and 0.25 square kilometres/year for the western coast.

Most of the eroded areas increase with larger areas of shrimp farms, less mangrove forest area, and when dams reduce riverine inputs and coastal land subsidence transpires. In areas where erosion has prevailed, the presence of mangroves has reduced erosion rates. Mangroves dominating coastal locations exhibit less erosion than areas with non-vegetated land or former mangrove areas.

Such examples indicate that there is a strong relation between major coastal erosion problems throughout the region and degradation of the protective function of coastal systems such as coastal forest and trees – particularly mangrove forest. Artificial and natural agents that induce mangrove loss and make coastal areas more susceptible to coastal erosion include anthropogenic factors such as excessive logging, direct land reclamation for agriculture, aquaculture, salt ponds, urban development and settlement, and to a lesser extent fires, storms, hurricanes, tidal waves and erosion cycles owing to changing sea levels.

More scientific investigation and quantification of the physical processes and dynamic interaction of the system is needed to understand how and under what circumstances mangrove forests and other coastal vegetation effectively protect the shoreline from erosion. A number of efforts have focused on field observations, laboratory and numerical model experiments and theoretical analysis

The Causes of Coastal Erosion

Coastal erosion and accretion are complex processes that need to be investigated from the angles of sediment motion under wind, wave and tidal current action; beach dynamics within a sediment/littoral cell; and human activities along the coast, within river catchments and watersheds and offshore, both at spatial and temporal scales. In terms of temporal scales, the issue of sea-level rise is complex and produces a range of environmental problems.

As the sea level rises, the water depth increases and the wave base becomes deeper; waves reaching the coast have more energy and therefore can erode and transport greater quantities of sediment. Thus, the coast starts to adjust to the new sea level to maintain a dynamic equilibrium.

Estuary

An estuary is a partly enclosed coastal body of brackish water with one or more rivers or streams flowing into it, and with a free connection to the open sea.

Estuaries form a transition zone between river environments and maritime environments and are subject to both marine influences—such as tides, waves, and the influx of saline water—and riverine influences, such as flows of fresh water and sediment. The inflows of both sea water and fresh water provide high levels of nutrients in both the water column and sediment, making estuaries among the most productive natural habitats in the world.

Most existing estuaries were formed during the Holocene epoch by the flooding of river-eroded or glacially scoured valleys when the sea level began to rise about 10,000-12,000 years ago. Estuaries are typically classified by their geomorphological features or by water circulation patterns and can be referred to by many different names, such as bays, harbors, lagoons, inlets, or sounds, although some of these water bodies do not strictly meet the above definition of an estuary and may be fully saline.

The banks of many estuaries are amongst the most heavily populated areas of the world, with about 60% of the world's population living along estuaries and the coast. As a result, many estuaries are suffering degradation by many factors, including sedimentation from soil erosion from deforestation, overgrazing, and other poor farming practices; overfishing; drainage and filling of wetlands; eutrophication due to excessive nutrients from sewage and animal wastes; pollutants including heavy metals, polychlorinated biphenyls, radionuclides and hydrocarbons from sewage inputs; and diking or damming for flood control or water diversion.

Classification based on Geomorphology

Drowned River Valleys: Their width-to-depth ratio is typically large, appearing wedge-shaped in the inner part and broadening and deepening seaward. Water depths rarely exceed 30 m (100 ft). Examples of this type of estuary in the U.S. are the Hudson River, Chesapeake Bay, and Delaware Bay along the Mid-Atlantic coast; and along the Gulf coast, Galveston Bay and Tampa Bay. San Francisco Bay is another good example of a drowned river valley.

Lagoon-type or bar-built: These estuaries are semi-isolated from ocean waters by barrier beaches (barrier islands and barrier spits). Formation of barrier beaches partially encloses the estuary, with only narrow inlets allowing contact with the ocean waters. Bar-built estuaries typically develop on gently sloping plains located along tectonically stable edges of continents and marginal sea coasts. They are extensive along the Atlantic and Gulf coasts of the U.S. in areas with active coastal deposition of sediments and where tidal ranges are less than 4 m (13 ft). The barrier beaches that enclose bar-built estuaries have been developed in several ways:

- building up of offshore bars by wave action, in which sand from the sea floor is deposited in elongated bars parallel to the shoreline,
- reworking of sediment discharge from rivers by wave, current, and wind action into beaches, overwash flats, and dunes,
- engulfment of mainland beach ridges (ridges developed from the erosion of coastal plain sediments around 5000 years ago) due to sea level rise and resulting in the breaching of the ridges and flooding of the coastal lowlands, forming shallow lagoons, and
- elongation of barrier spits from the erosion of headlands due to the action of longshore currents, with the spits growing in the direction of the littoral drift.

Barrier beaches form in shallow water and are generally parallel to the shoreline, resulting in long, narrow estuaries. The average water depth is usually less than 5 m (16 ft), and rarely exceeds 10 m (33 ft). Examples of bar-built estuaries are Barnegat Bay, New Jersey; Laguna Madre, Texas; and Pamlico Sound, North Carolina.

Fjord-type

Fjord-type estuaries are formed in deeply eroded valleys formed by glaciers. These U-shaped estuaries typically have steep sides, rock bottoms, and underwater sills contoured by glacial movement. The estuary is shallowest at its mouth, where terminal glacial moraines or rock bars form sills that restrict water flow. In the upper reaches of the estuary, the depth can exceed 300 m (1,000 ft). The width-to-depth ratio is generally small. In estuaries with very shallow sills, tidal oscillations only affect the water down to the depth of the sill, and the waters deeper than that may remain stagnant for a very long time, so there is only an occasional exchange of the deep water of the estuary with the ocean. If the sill depth is deep, water circulation is less restricted, and there is a slow but steady exchange of water between the estuary and the ocean. Fjord-type estuaries can be found along the coasts of Alaska, the Puget Sound region of western Washington state,

British Columbia, eastern Canada, Greenland, Iceland, New Zealand, and Norway.

Tectonically Produced

These estuaries are formed by subsidence or land cut off from the ocean by land movement associated with faulting, volcanoes, and landslides. Inundation from eustatic sea level rise during the Holocene Epoch has also contributed to the formation of these estuaries. There are only a small number of tectonically produced estuaries; one example is the San Francisco Bay, which was formed by the crustal movements of the San Andreas fault system causing the inundation of the lower reaches of the Sacramento and San Joaquin rivers.

Classification based on Water Circulation

Salt wedge: In this type of estuary, river output greatly exceeds marine input and tidal effects have a minor importance. Fresh water floats on top of the seawater in a layer that gradually thins as it moves seaward. The denser seawater moves landward along the bottom of the estuary, forming a wedge-shaped layer that is thinner as it approaches land. As a velocity difference develops between the two layers, shear forces generate internal waves at the interface, mixing the seawater upward with the freshwater. An example of a salt wedge estuary is the Mississippi River.

Partially Mixed

As tidal forcing increases, river output becomes less than the marine input. Here, current induced turbulence causes mixing of the whole water column such that salinity varies more longitudinally rather than vertically, leading to a moderately stratified condition. Examples include the Chesapeake Bay and Narragansett Bay.

Vertically Homogenous

Tidal mixing forces exceed river output, resulting in a well mixed water column and the disappearance of the vertical salinity gradient. The freshwater-seawater boundary is eliminated due to the intense turbulent mixing and eddy effects. The lower reaches of Delaware Bay and the Raritan River in New Jersey are examples of vertically homogenous estuaries.

Inverse

Inverse estuaries occur in dry climates where evaporation greatly exceeds the inflow of fresh water. A salinity maximum zone is formed, and both riverine and oceanic water flow close to the surface towards this zone. This water is pushed downward and spreads along the bottom in both the seaward and landward direction. An example of an inverse estuary is Spencer Gulf, South Australia.

Intermittent

Estuary type varies dramatically depending on freshwater input, and is capable of changing from a wholly marine embayment to any of the other estuary types.Estuarine water circulation

Estuarine water circulation is controlled by the inflow of rivers, the tides, rainfall and evaporation, the wind, and other oceanic events such as an upwelling, an eddy, and storms. Water circulation patterns are influenced by vertical mixing and stratification, and can affect residence time and exposure time.

The residence time of water is a key variable determining the health of an estuary, particularly from human-induced stresses. Rapid flushing ensures that there is insufficient time for sediment accumulation or dissolved oxygen depletion in the estuary; thus a well flushed estuary is intrinsically more robust than a poorly flushed estuary. Residence time also affects other parameters such as heavy metals, dissolved nutrients, suspended solids, and algal blooms that may affect the health of estuaries.

A simple way to calculate residence time is using a simple classic estuary model, which can be useful for acquiring a conceptual understanding of an estuary, but is coarse in time and space. A classic estuary has the following components: 1) freshwater inflow with a discharge Q_f and a salinity S_f (generally $S_f = 0$); 2) oceanic inflow with a discharge Q_{in} and salinity S_o; and 3) outflow to the ocean with a discharge of Q_{out} and a salinity S_1. The inflow and outflow of water is equal because mass is conserved. Salt is also a conserved, therefore, the inflow and outflow fluxes of salinity are also equal. If groundwater inputs and evaporation are ignored, the continuity equation is:

$$Q_{in} S_o + Q_f S_f = Q_{out} S_1$$

The residence time T is the volume of water within the estuary (Vol) divided by the flow rate of the river:

$$T = (Vol/ Q_f)(1 - S_1/ S_o - S_f / S_o)$$

Exposure Time

Residence time considers the time it takes for the water particles to leave the estuary, however, some water particles that leave the estuary during an ebb tide may re-enter the system during a flood tide. The amount of time a water particle spends in the estuary until it never returns is called exposure time. The exposure time can be much larger than the residence time if the water particles are leaving with the ebb tides and returning with the rising tide. The ratio between the number of water particles returning to the estuary and the number of water particles leaving is known as the return coefficient, r.

In order to quantify exposure time, the water circulation outside of the estuary must be determined. However, the circulation near the mouth of the estuary is complex due to the tidal mixing processes that occur between the estuarine and ocean waters. If the coast is rugged with headlands, a mosaic of complex flow fields consisting of eddies, jets, and stagnation zones will occur, further complicating the circulation patterns outside of the estuary.

In cases involving deltas or wetlands that drain into multiple tidal creeks, such as Missionary Bay, Australia, water leaving one creek at the ebb tide may enter another estuary during the flood tide. When there is a series of estuaries involved, a large exposure time (larger than that of the individual estuaries), will occur if the tidal outflow from one estuary re-enters a different estuary during the flood tide. Along a rugged coastline with headlands, however, mixing of estuary and oceanic waters can be intense. When estuarine water leaves the estuary, it gets flushed out to coastal waters, so exposure time and residence time are close equal.

In some cases it is possible to measure the fluxes of volume, salt, and temperature across the mouth of an estuary through a tidal cycle. Using this data, (1-r) can be calculated (r is the return coefficient). (1-r) is equal to the fraction of the volume of water V_{TP} (mean tidal prism volume) leaving the estuary during the ebb tide that is replaced with coastal waters prior to re-entering the system. When r = 1, the same water is re-entering the estuary, and if r = 0, the estuarine water that has left the estuary during the ebb tide has been replaced with coastal waters entering the estuary during the rising tide. The exposure time τ is estimated by:

$$\tau = V_{estuary}\, T_{tide} \,/\, (1\text{-}r)\, V_{TP}$$

$V_{estuary}$ is defined as the mean estuarine volume and T_{tide} is the tidal period.

The total fluxes of brackish water through the river mouth during tidal events is often much higher (often by a factor of 10 to 100) than the volume flux from riverine inflow. Therefore, if measurements are not precise, the estimate of the net flux will be unreliable. Direct measurements of the return coefficient are often complicated by unsteady oceanic events such an upwelling, the passage of an eddy, or storms, so the success of a correct direct measurement of the return coefficient is rare.

Vertical Mixing and Stratification

The residence time of water in an estuary is dependent on the circulation within the estuary that is driven by density differences due to changes in salinity and temperature. Less dense freshwater floats over saline water and warmer water floats above colder water (temperatures greater than 4°C). As a result, near-surface and near-bottom waters can have different trajectories, resulting in different residence times.

Vertical mixing determines how much the salinity and temperature will change from the top to the bottom, profoundly affecting water circulation. Vertical mixing occurs at three levels: from the surface downward by wind forces, the bottom upward by boundary generated turbulence (estuarine and oceanic boundary mixing), and internally by turbulent mixing caused by the water currents which are driven by the tides, wind, and river inflow.

Different Types of Estuarine Circulation Result from Vertical Mixing

Salt Wedge Estuary

These estuaries are characterized by a sharp density interface between the upper layer of freshwater and the bottom layer of saline water. River water dominates in this system, and tidal effects have a small role in the circulation patterns. The freshwater floats on top of the seawater and gradually thins as it moves seaward. The denser seawater moves along the bottom up the estuary forming a wedge shaped layer and becoming thinner as it moves landward. As a velocity difference develops between the two layers, shear forces generate internal waves at the interface, mixing the seawater upward with the freshwater. An example is the Mississippi estuary.

Partially Stratified Estuary

As tidal forcing increases, the control of river flow on the pattern of circulation in the estuary becomes less dominating. Turbulent mixing induced by the current creates a moderately stratified condition. Turbulent eddies mix the water column, creating a mass transfer of freshwater and seawater in both directions across the density boundary. Therefore, the interface separating the upper and lower water masses is replaced with a water column with a gradual increase in salinity from surface to bottom. A two layered flow still exists however, with the maximum salinity gradient at mid depth. Partially stratified estuaries are typically shallow and wide, with a greater width to depth ratio than salt wedge estuaries. An example is the Thames.

Vertically Homogeneous Estuary

In these estuaries, tidal flow is greater relative to river discharge, resulting in a well mixed water column and the disappearance of the vertical salinity gradient. The freshwater-seawater boundary is eliminated due to the intense turbulent mixing and eddy effects. The width to depth ratio of vertically homogeneous estuaries is large, with the limited depth creating enough vertical shearing on the seafloor to mix the water column completely. If tidal currents at the mouth of an estuary are strong enough to create turbulent mixing, vertically homogeneous conditions often develop.

Fjords

Fjords are examples of highly stratified estuaries; they are silled basins that have freshwater inflow that greatly exceeds evaporation. Oceanic water is imported in an intermediate layer and mixes with the freshwater. The resulting brackish water is then exported into the surface layer. A slow import of seawater may flow over the sill and sink to the bottom of the fjord (deep layer), where the water remains stagnant until flushed by an occasional storm.

Inverse Estuary

Inverse estuaries occur in dry climates where evaporation greatly exceeds the inflow of freshwater. A salinity maximum zone is formed, and both riverine and oceanic water flow close to the surface towards this zone. This water is pushed downward and spreads along the bottom in both the seaward and landward direction. The maximum salinity can reach extremely high values and the residence time can be several months. In these systems, the salinity maximum zone acts like a plug, inhibiting the mixing of estuarine and oceanic waters so that freshwater does not reach the ocean. The high salinity water sinks seaward and exits the estuary.

Physiochemical Variation

The most important variable characteristics of estuary water are the concentration of dissolved oxygen, salinity and sediment load. There is extreme spatial variability in salinity, with a range of near zero at the tidal limit of the tributary river(s) to 3.4% at the estuary mouth. At any one point the salinity will vary considerably over time and seasons, making it a harsh environment for organisms. Sediment often settles in intertidal mudflats which are extremely difficult to colonize. No points of attachment exist for algae, so vegetation based habitat is not established. Sediment can also clog feeding and respiratory structures of species, and special adaptations exist within mudflat species to cope with this problem. Lastly, dissolved oxygen variation can cause problems for life forms. Nutrient-rich sediment from man-made sources can promote primary production life cycles, perhaps leading to eventual decay removing the dissolved oxygen from the water; thus hypoxic or anoxic zones can develop.

Implications for Marine Life

Estuaries provide habitats for a large number of organisms and support very high productivity. Estuaries provide habitats for many fish nurseries, depending upon their locations in the world, such as salmon and sea trout. Also, migratory bird populations, such as the black-tailed godwit, *Limosa limosa islandica* make essential use of estuaries.

Two of the main challenges of estuarine life are the variability in salinity and sedimentation. Many species of fish and invertebrates have various methods to control or conform to the shifts in salt concentrations and are termed osmoconformers and osmoregulators. Many animals also burrow to avoid predation and to live in the more stable sedimental environment. However, large numbers of bacteria are found within the sediment which have a very high oxygen demand. This reduces the levels of oxygen within the sediment often resulting in partially anoxic conditions, which can be further exacerbated by limited water flux.

Phytoplankton are key primary producers in estuaries. They move with the water bodies and can be flushed in and out with the tides. Their productivity is largely dependent upon the turbidity of the water. The main phytoplankton present are diatoms and dinoflagellates which are abundant in the sediment.

It is important to remember that a primary source of food for many organisms on estuaries, including bacteria, is detritus from the settlement of the sedimentation.

Human Impact

Of the thirty-two largest cities in the world, twenty-two are located on estuaries. For example, New York City is located at the mouth of the Hudson River estuary.

As ecosystems, estuaries are under threat from human activities such as pollution and overfishing. They are also threatened by sewage, coastal settlement, land clearance and much more. Estuaries are affected by events far upstream, and concentrate materials such as pollutants and sediments. Land run-off and industrial, agricultural, and domestic waste enter rivers and are discharged into estuaries. Contaminants can be introduced which do not disintegrate rapidly in the marine environment, such as plastics, pesticides, furans, dioxins, phenols and heavy metals.

Such toxins can accumulate in the tissues of many species of aquatic life in a process called bioaccumulation. They also accumulate in benthic environments, such as estuaries and bay muds: a geological record of human activities of the last century. For example, Chinese and Russian industrial pollution, such as phenols and heavy metals, has devastated fish stocks in the Amur River and damaged its estuary soil.

Estuaries tend to be naturally eutrophic because land runoff discharges nutrients into estuaries. With human activities, land run-off also now includes the many chemicals used as fertilizers in agriculture as well as waste from livestock and humans. Excess oxygen-depleting chemicals in the water can lead to hypoxia and the creation of dead zones. This can result in reductions in water quality, fish, and other animal populations. Overfishing also occurs, Chesapeake Bay once had a flourishing oyster population which has been almost wiped out by overfishing. Oysters filter these pollutants, and either eat them or shape them into small packets that are deposited on the bottom where they are harmless. Historically the oysters filtered the estuary's entire water volume of excess nutrients every three or four days. Today that process takes almost a year, and sediment, nutrients, and algae can cause problems in local waters.

Coastal Landform Changes

Coastal landforms, any of the relief features present along any coast, the result of a combination of processes, sediments, and the geology of the coast itself.

The coastal environment of the world is made up of a wide variety of landforms manifested in a spectrum of sizes and shapes ranging from gently sloping beaches to high cliffs, yet coastal landforms are best considered in two broad categories: erosional and depositional. In fact, the overall nature of any coast may be described in terms of one or the other of these categories. It should be noted, however, that each of the two major landform types may occur on any given reach of coast.

A number of mechanical and chemical effects produce erosion of rocky shorelines by waves. Depending on the geology of the coastline, nature of wave attack, and long-term changes in sea-level as well as tidal ranges, erosional landforms such as wave-cut notches, sea cliffs and even unusual landforms such as caves, sea arches, and sea stacks can form. Transportation by waves and currents is necessary in order to move rock particles eroded from one part of a coastline to a place of deposition elsewhere.

One of the most important transport mechanisms results from wave refraction. Since waves rarely break onto a shore at right angles, the upward movement of water onto the beach (swash) occurs at an oblique angle. However, the return of water (backwash) is at right angles to the beach, resulting in the net movement of beach material laterally. This movement is known as beach drift. The endless cycle of swash and backwash and resulting beach drift can be observed on all beaches.

When incoming waves strike a shoreline, beach drift can cause the movement of sediment along the shoreline. Beach drift begins when the waves reach the shore at an angle (not parallel) causing the swash and entrained sediment to travel obliquely up the beach. The returning backwash moves back to the ocean with a direction that is defined by the slope of the beach. The differences in the direction of the swash and backwash cause a net longshore movement of sediment.

Frequently, backwash and rip currents cannot remove water from the shore zone as fast as it is piled up there by waves. As a result, there is a buildup of water that results in the lateral movement of water and sediment just offshore in a direction with the waves. The currents produced by the laterial movement of water are known as longshore currents. The movement of sediment is known as longshore drift, which is distinct from the beach drift described earlier which operates on land at the beach. The combined movement of sediment via longshore drift and beach drift is known as littoral drift.

Tidal currents along coasts can also be effective in moving eroded material. While incoming and outgoing tides produce currents in opposite directions on a daily basis, the current in one direction is usually stronger than in the other resulting in a net one-way transport of sediment. Longshore drift, longshore currents, and tidal currents in combination determine the net direction of sediment transport and areas of deposition.

Many kinds of depositional landforms are possible along coasts depending on the configuration of the original coastline, direction of sediment transport, nature of the waves, and shape and steepness of the offshore underwater slope. Some common depositional forms are spits, bayhead beaches, barrier beaches or bay-mouth bars, tombolos, and cuspate forelands.

Factors and Forces in the Formation of Coastal Features

The landforms that develop and persist along the coast are the result of a combination of processes acting upon the sediments and rocks present in the coastal zone. The most prominent of these processes involves waves and the currents that they generate, along with tides. Other factors that significantly affect coastal morphology are climate and gravity.

Waves

The most obvious of all coastal processes is the continual motion of the waves moving toward the beach. Waves vary considerably in size over time at any given location and also vary markedly from place to place. Waves interact with the ocean bottom as they travel into shallow water; as a result, they cause sediment to become temporarily suspended and available for movement by coastal currents. The larger the wave, the deeper the water in which this process takes place and the larger the particle that can be moved. Even small waves that are only a few tens of centimetres high can pick up sand as they reach the shore. Larger waves can move cobbles and rock material as large as boulders.

Generally, small waves cause sediment—usually sand—to be transported toward the coast and to become deposited on the beach. Larger waves, typically during storms, are responsible for the removal of sediment from the coast and its conveyance out into relatively deep water.

Waves erode the bedrock along the coast largely by abrasion. The suspended sediment particles in waves, especially pebbles and larger rock debris, have much the same effect on a surface as sandpaper does.

Waves have considerable force and so may break up bedrock simply by impact.

Longshore Currents

Waves usually approach the coast at some acute angle rather than exactly parallel to it. Because of this, the waves are bent (or refracted) as they enter shallow water, which in turn generates a current along the shore and parallel to it. Such a current is called a longshore current, and it extends from the shoreline out through the zone of breaking waves. The speed of the current is related to the size of the waves and to their angle of approach. Under rather quiescent conditions, longshore currents move only about 10–30 centimetres per second; however, under stormy conditions they may exceed one metre per second. The combination of waves and longshore current acts to transport large quantities of sediment along the shallow zone adjacent to the shoreline.

Because longshore currents are caused by the approaching and refracting waves, they may move in either direction along the coast, depending on the direction of wave approach. This direction of approach is a result of the wind direction, which is therefore the ultimate factor in determining the direction of longshore currents and the transport of sediment along the shoreline.

Although a longshore current can entrain sediment if it moves fast enough, waves typically cause sediment to be picked up from the bottom, and the longshore current transports it along the coast. In some locations there is quite a large volume of net sediment transport along the coast because of a dominance of one wind direction—and therefore wave direction—over another. This volume may be on the order of 100,000 cubic metres per year. Other locations may experience more of a balance in wave approach, which causes the longshore current and sediment transport in one direction to be nearly balanced by the same process in the other direction.

Rip Currents

Another type of coastal current caused by wave activity is the rip current (incorrectly called rip tide in popular usage). As waves move toward the beach, there is some net shoreward transport of water. This leads to a slight but important upward slope of the water level (setup), so that the absolute water level at the shoreline is a few centimetres higher than it is beyond the surf zone. This situation is an unstable one, and water moves seaward through the surf zone in an effort to relieve the instability of the sloping water. The seaward movement is typically confined to narrow pathways. In most cases, rip currents are regularly spaced and flow at speeds of up to several tens of centimetres per second. They can carry sediment and often are recognized by the plume of suspended sediment moving out through the surf zone. In some localities rip currents persist for months at the same site, whereas in others they are quite ephemeral.

Spit

A spit or sandspit is a deposition landform found off coasts. At one end, spits connect to a head, and extend into the nose. A spit is a type of bar or beach that develops where a re-entrant occurs, such as at cove's headlands, by the process of longshore drift. Longshore drift (also called littoral drift) occurs due to waves meeting the beach at an oblique angle, and backwashing perpendicular to the shore, moving sediment down the beach in a zigzag pattern. Longshore drifting is complemented by longshore currents, which transport sediment through the water alongside the beach. These currents are set in motion by the same oblique angle of entering waves that causes littoral drift and transport sediment in a similar process.

Hydrology and Geology

Where the direction of the shore inland *re-enters*, or changes direction, such as at a headland, the longshore current spreads out or dissipates. No longer able to carry the full load, much of the sediment is dropped, called deposition. This submerged bar of sediment allows longshore drift or littoral drift, to continue to transport in the direction the waves are breaking, forming an above-water spit. Without the complementary process of littoral drift, the bar would not build above the surface of the waves becoming a spit and would instead be leveled off underwater.

Spits occur when longshore drift reaches a section of headland where the turn is greater than 30 degrees. They will continue out into the sea until water pressure (such as from a river) becomes too much to allow the sand to deposit. The spit may then be grown upon and become stable and often fertile. A spit may be considered a specialized form of a *shoal*. As spits grow, the water behind them is sheltered from wind and waves, and a salt marsh is likely to develop.

Wave refraction can occur at the end of a spit, carrying sediment around the end to form a hook or recurved spit. Wave refraction in multiple directions will cause a complex spit to form. Incoming waves that come in a direction other than obliquely along the spit will halt the growth of the spit, shorten it or eventually destroy it entirely.

The sediments that make up spits come from a variety of sources including rivers and eroding bluffs, and changes there can have a large impact on spits and other coastal landforms. Activities such as logging and farming upstream can increase the sediment load of rivers, which may hurt the intertidal environments around spits by smothering delicate habitat. Roads or bulkheads built along bluffs can drastically reduce the volume of sediment eroded, so that not enough material is being pushed along to maintain a given spit.

If the supply of sediment is interrupted the sand at the neck (landward side) of the spit may be moved towards the head, eventually creating an island. If the supply isn't interrupted, and the spit isn't breached by the sea (or, if across an estuary, the river) the spit may become a bar, with both ends joined to land, and form a lagoon behind the bar. If an island lies offshore near where the coast changes direction, and the spit continues to grow until it connects the island to the mainland, it is then called a tombolo.

The end of a spit attached to land is called the proximal end, and the end jutting out into water is called the distal end.

Spits around the world

The longest spit in the world is the Arabat Spit in the Sea of Azov. It is approximately 110 km (68 mi) long.

Farewell Spit in New Zealand, at 32 km (20 mi), in the North West corner of South Island is believed to be caused by the strong prevailing winds and currents bringing sand eroded from the Southern Alps of the South Island and depositing these into Golden Bay.

Spits in the UK are caused by prevailing South-Westerly winds, which give the spits their direction. However, when the direction of the wind changes for a short while the spit may change in direction for a short while forming a hook. Many spits have hooked or curved ends. One spit in the UK can be found in Dorset. Chesil Beach is a 29 km (18 mi) long shingle that connects Weymouth to the Isle of Portland. Chesil Beach provides shelter to Weymouth and the Portland village of Chiswell from prevailing winds and waves. You can also find Spurn Point in Yorkshire and the Humber it is approximately 4.8 km (3.0 mi).

The Curonian Spit off the coast of Lithuania and Kaliningrad Oblast of Russia separates Curonian Lagoon from the Baltic Sea and is 98 km (61 mi) long. In a similar fashion, the Vistula Spit separates Vistula Lagoon from the Gdańsk Bay off the coast of Poland.

Human Settlement Patterns

Since prehistory humans have chosen certain spit formations as sites for human habitation. In some cases such as Chumash Native American prehistorical settlement, these sites have been chosen for proximity to marine resource exploitation; for example, the Chumash settlement on the Morro Bay sandspit is one such location.

Delineation of Submerged Paleoshorelines and Paleochannels

Submergent coastlines are stretches along the coast that have been inundated by the sea due to a relative rise in sea levels. This occurs due to either isostacy or eustacy.

Submergent coastline are the opposite of emergent coastlines which have experienced a relative fall in sea-levels.

Features of a Submergent Coastline are

- drowned river valleys or rias
- drowned glaciated valleys or fjords.

Estuaries are often the Drowned Mouths of Rivers

The Western Coastal Plains of the Indian subcontinent are examples of submergent coastline. An ancient city of Dvârakâ, mentioned in the great epic Mahabharata, is under water now. The coastline also forms the estuaries of the Narmada and Tapti rivers.

Few things are as temporary as footprints on a sandy beach. They are unmistakably human and fleetingly ephemeral—etched in passing and erased with the next wave or high tide. But as with footprints, the steady pounding of ocean waves can reduce built shorelines, roads, cities and civilizations to rubble—a constant reminder of the energy and restlessness of the seas. Oceans, never at rest, transgress the shore with each wave, each tide cycle, and in repeating intervals measured in centuries and millenia.

During the life of our planet, great changes in sea level have occurred. Fossil shells tumble out of mountaintops, and eroded plains appear underwater. In these is the evidence of the ongoing dance between the continents and the oceans, the material cycles of uplift and erosion, and the changing shape of the shores.

We know that in the last 15,000 years—the generally-accepted era of human occupancy of North American coastlines—sea level has varied from more than 100 meters below to as high as 10 meters above the present sea level. Much of this change is the result of climate. When North America was gripped in an Ice Age, water was confined to vast glacial systems resting atop land masses. As the climate warmed, glaciers melted, literally over-filling the oceans and causing the sea level to rise. Other processes affect shorelines as well. "Isostatic rebound" represents the movement of land masses in response to the massive weight of continental glaciers. As glaciers melt and sea level rises, the land rises slightly, unburdened by the ice load.

Traditional scientific accounts of human migration to North America are based heavily on the dynamics of sea-level change. When the Bering Land Bridge was exposed, a thousand-mile-wide expanse of green land lay between Asia and North America—a welcoming carpet for ancestors of Native Americans. The conventional wisdom of the land-bridge migration story has held that humans found their way southward into North America via ice-free corridors. They moved quickly as game-hunting generalists who followed their quarry into the grasslands, open valleys and plains in the central part of the continent. Ocean-based cultures, according to this explanation, developed more recently, as people specialized and adapted to maritime climates and marine ecological resources.

Yet, archaeological evidence from North and South America suggests that humans have been in the Americas for perhaps longer than the "ice-free corridor" can safely explain. Estimates of human settlements as old as 30,000 years represent serious challenges to traditional accounts.

Recently, another explanation has gained favor—that prehistoric emigrants from Asia arrived by way of the coastlines, as maritime people. If so, then coastal regions—and particularly areas submerged since sea level rise—may bear the archaeological evidence of human occupancy.

The study of paleoshorelines ("paleo" meaning old or ancient) examines the complex intersection of geology, climate, ocean research and archaeology. Scientists must first find ancient coastlines, rivers, estuaries, and floodplains, and then look for signs of people—tools, firepits, middens (refuse heaps, often shells and bones), and remnants or signs of habitations.

Locating the landforms where humans once lived is difficult, because they are now under water. Assuming that people fished and hunted, and lived in places not too different from the places that they occupied in more recent prehistory, then the submerged landscapes form a starting point for underwater search.

Palaeochannel

A palaeochannel or paleochannel is a remnant of an inactive river or stream channel that has been either filled or buried by younger sediment. The sediments that the ancient channel is either cut into or buried by can be either unconsolidated, semi-consolidated, consolidated, or lithified. The word palaeochannel is formed from the words "palaeo" or 'old', and channel; ie; a palaeochannel is an old channel. This may be synonymous with palaeovalley and palaeoriver.

A palaeochannel is distinct from the overbank deposits of currently active river channels, including ephemeral water courses which do not regularly flow (such as the Todd River, Central Australia). A palaeochannel is distinct from such watercourses because the river bed is filled with sedimentary deposits which are unrelated to the normal bed load of the current drainage pattern.

Many palaeochannels are arranged on old drainage patterns which are distinct from the current drainage system of a catchment. For example, palaeochannels may relate to a system of rivers and creeks which drained east-west and the current drainage direction is north-south.

Palaeochannels can be most easily identified as broad erosional channels into a basement which underlies a system of depositional sequences which may contain several episodes of deposition and represent meandering peneplain streams.

Thereafter, a palaeochannel may form part of the regolith of a region and, although it is unconsolidated or partly consolidated, is currently part of the erosional surface.

Palaeochannels can also be identified according to their age. For example, there are deposits of Tertiary lignites situated in Tertiary river systems which are preserved on top of Archaean basement in the Yilgarn Craton of Western Australia. These river systems have laid in place for ~15 to 50 million years and would be considered palaeochannels.

Formation

Paleochannels form when river channels aggrade, depositing sediment on their bed. For these channel deposits to be preserved, the flow must not re-occupy and re-erode them. Examples of what may cause long-term preservation include the channels being in a net-depositional environment and/or being in a subsiding sedimentary basin. Paleochannels may also be preserved in the short-term on non-net-depositional floodplains in which the river migrates or avulses away from its previous course. This preservation is short-term because unless the channel deposits are buried, flow will eventually re-occupy its formerly-occupied course, re-working and eroding the channel deposits.

Geological Importance

Palaeochannels are important to geology for a number of reasons;

- Understanding movements of faults, which may redirect river systems and thus form stranded channels which are in essence palaeochannels.
- Preserving Tertiary, Eocene and Recent sediments and fossils within them, and are important locations for palaeontology, palaeobotany and archaeology.
- Preserving evidence of older erosional surfaces and levels, useful for estimating the net erosional budget of older regolith
- Preserving sedimentary records useful for understanding climatic conditions, including various isotopic indicators of past rainfall, temperatures and climates, which are used in understanding climate change and global warming

Economic Importance

Palaeochannels can host economic ore deposits of uranium, lignite, precious metals such as gold and platinum, heavy minerals such as tin, tungsten, and iron ore preserved as paleo-placer deposits.

Effect of Hydrometeorological Conditions on Sediment

Sediment is a naturally occurring material that is broken down by processes of weathering and erosion, and is subsequently transported by the action of wind, water, or ice, and/or by the force of gravity acting on the particles. For example, silt falls out of suspension via sedimentation and forms soil (some of which may eventually become sedimentary rock).

Sediments are most often transported by water (fluvial processes), wind (aeolian processes) and glaciers. Beach sands and river channel deposits are examples of fluvial transport and deposition, though sediment also often settles out of slow-moving or standing water in lakes and oceans. Desert sand dunes and loess are examples of aeolian transport and deposition. Glacial moraine deposits and till are ice-transported sediments.

Classification

Sediment can be classified based on its grain size and/or its composition.

Grain Size

Sediment size is measured on a log base 2 scale, called the "Phi" scale, which classifies particles by size from "colloid" to "boulder".

φ scale	*Size range (metric)*	*Size range (inches)*	*Aggregate class (Wentworth)*	*Other names*
< –8	> 256 mm	> 10.1 in	Boulder	
–6 to –8	64–256 mm	2.5–10.1 in	Cobble	
–5 to –6	32–64 mm	1.26–2.5 in	Very coarse gravel	Pebble
–4 to –5	16–32 mm	0.63–1.26 in	Coarse gravel	Pebble
–3 to –4	8–16 mm	0.31–0.63 in	Medium gravel	Pebble
–2 to –3	4–8 mm	0.157–0.31 in	Fine gravel	Pebble
–1 to –2	2–4 mm	0.079–0.157 in	Very fine gravel	Granule
0 to –1	1–2 mm	0.039–0.079 in	Very coarse sand	
1 to 0	0.5–1 mm	0.020–0.039 in	Coarse sand	
2 to 1	0.25–0.5 mm	0.010–0.020 in	Medium sand	
3 to 2	125–250 μm	0.0049–0.010 in	Fine sand	
4 to 3	62.5–125 μm	0.0025–0.0049 in	Very fine sand	
8 to 4	3.9–62.5 μm	0.00015–0.0025 in	Silt	Mud
> 8	< 3.9 μm	< 0.00015 in	Clay	Mud
>10	< 1 μm	< 0.000039 in	Colloid	Mud

Composition

Composition of sediment can be measured in terms of:

- parent rock lithology
- mineral composition
- chemical make-up.

This leads to an ambiguity in which clay can be used as both a size-range and a composition.

Sediment Transport

Sediment is transported based on the strength of the flow that carries it and its own size, volume, density, and shape. Stronger flows will increase the lift and drag on the particle, causing it to rise, while larger or denser particles will be more likely to fall through the flow.

Fluvial Processes: Rivers, Streams, and Overland Flow

Particle motion: Rivers and streams carry sediment in their flows. This sediment can be in a variety of locations within the flow, depending on the balance between the upwards velocity on the particle (drag and lift forces), and the settling velocity of the particle. These relationships are shown in the following table for the Rouse number, which is a ratio of sediment fall velocity to upwards velocity.

$$\text{Rouse} = \frac{\text{Settling velocity}}{\text{Upwards velocity from lift and drag}} = \frac{w_s}{\kappa u_*}$$

where

- w_s is the fall velocity
- $k$is the von Kármán constant
- u_* is the shear velocity

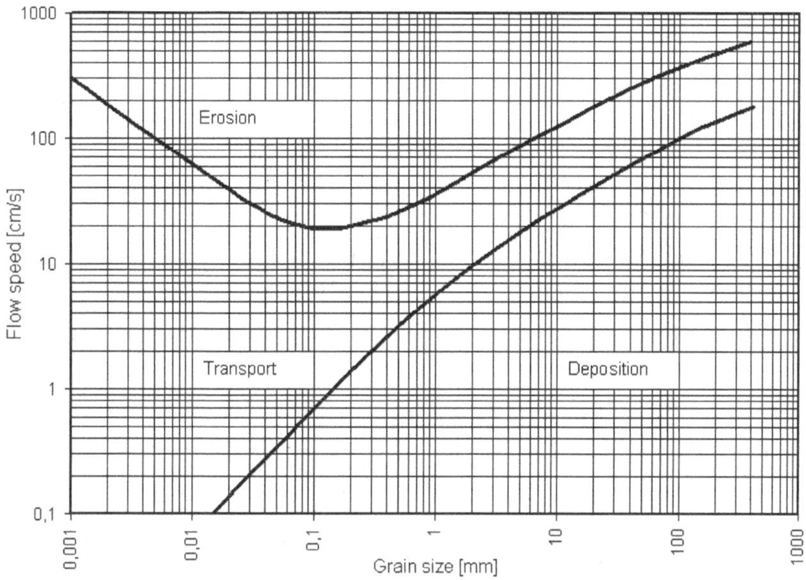

Figure: *Hjulström curve: The velocities of currents required for erosion, transportation, and deposition (sedimentation) of sediment particles of different sizes.*

Mode of Transport	***Rouse Number***
Bed load	>2.5
Suspended load: 50% Suspended	>1.2, <2.5
Suspended load: 100% Suspended	>0.8, <1.2
Wash load	<0.8

If the upwards velocity approximately equal to the settling velocity, sediment will be transported downstream entirely as suspended load. If the upwards velocity is much less than the settling velocity, but still high enough for the sediment to move, it will move along the bed as bed load by rolling, sliding, and saltating (jumping up into the flow, being transported a short distance then settling again). If the upwards velocity is higher than the settling velocity, the sediment will be transported high in the flow as wash load.

As there are generally a range of different particle sizes in the flow, it is common for material of different sizes to move through all areas of the flow for given stream conditions.

Fluvial Bedforms

Sediment motion can create self-organized structures such as ripples, dunes, antidunes on the river or stream bed. These bedforms are often preserved in sedimentary rocks and can be used to estimate the direction and magnitude of the flow that deposited the sediment.

Surface Runoff

Overland flow can erode soil particles and transport them downslope. The erosion associated with overland flow may occur through different methods depending on meteorological and flow conditions.

- If the initial impact of rain droplets dislodges soil, the phenomenon is called rainsplash erosion.
- If overland flow is directly responsible for sediment entrainment but does not form gullies, it is called "sheet erosion".
- If the flow and the substrate permit channelization, gullies may form; this is termed "gully erosion".

Key Fluvial Depositional Environments

The major fluvial (river and stream) environments for deposition of sediments include:

- Deltas (arguably an intermediate environment between fluvial and marine)
- Point bars
- Alluvial fans
- Braided rivers
- Oxbow lakes
- Levees
- Waterfalls

Aeolian Processes: Wind

Wind results in the transportation of fine sediment and the formation of sand dune fields and soils from airborne dust.

Glacial Processes

Glaciers carry a wide range of sediment sizes, and deposit it in moraines.

Mass Balance

The overall balance between sediment in transport and sediment being deposited on the bed is given by the Exner equation. This expression states that the rate of increase in bed elevation due to deposition is proportional to the amount of sediment that falls out of the flow. This equation is important in that changes in the power of the flow changes the ability of the flow to carry sediment, and this is reflected in patterns of erosion and deposition observed throughout a stream. This can be localized, and simply due to small obstacles: examples are scour holes behind boulders, where flow accelerates, and deposition on the inside of meander bends. Erosion and deposition can also be regional: erosion can occur due to dam removal and base level fall. Deposition can occur due to dam emplacement that causes the river to pool, and deposit its entire load or due to base level rise.

Shores and Shallow Seas

Seas, oceans and lakes accumulate sediment over time. The sediment could consist of *terrigenous* material, which originates on land, but may be deposited in either terrestrial, marine, or lacustrine (lake) environments; or of sediments (often biological) originating in the body of water. Terrigenous material is often supplied by nearby rivers and streams or reworked marine sediment (e.g. sand). In the mid-ocean, living organisms are primarily responsible for the sediment accumulation, their shells sinking to the ocean floor upon death.

Deposited sediments are the source of sedimentary rocks, which can contain fossils of the inhabitants of the body of water that were, upon death, covered by accumulating sediment. Lake bed sediments that have not solidified into rock can be used to determine past climatic conditions.

Key Marine Depositional Environments

The major areas for deposition of sediments in the marine environment include:

- Littoral sands (e.g. beach sands, runoff river sands, coastal bars and spits, largely clastic with little faunal content)
- The continental shelf (silty clays, increasing marine faunal content).
- The shelf margin (low terrigenous supply, mostly calcareous faunal skeletons)
- The shelf slope (much more fine-grained silts and clays)
- Beds of estuaries with the resultant deposits called "bay mud".

One other depositional environment which is a mixture of fluvial and marine is the turbidite system, which is a major source of sediment to the deep sedimentary and abyssal basins as well as the deep oceanic trenches.

Any depression in a marine environment where sediments accumulate over time is known as a sediment trap.

The null point theory explains how sediment deposition (geology) undergoes a hydrodynamic sorting process within the marine environment leading to a seaward fining of sediment grain size.

Application of Remote Sensing and Mike21 for Sediment

The Earth's surface undergoes constant changes and transformations under the influence of both, natural and man-made processes. The changes affect individual elements of the Earth's surface and entire landscape structures. Small river catchments are spatial units (geoecosystems) that are especially sensitive to natural and man-made processes. External impulses and their effects are immediately coded in a system and possible to register under the integrated monitoring of the natural environment. Small river catchments are a very important element of the contemporary landscape of the Earth's surface. Whether at the regional, continental or global scales, they are worth preserving and protecting.

GIS and Remote Sensing were demonstrated as tools for assessing the status of the mangroves, coral reefs, mapping of sea grass beds, biodiversity of the habitats, etc. Changes in the characteristics of mangrove ecosystem, especially the mangrove vegetation coverage and the land cover changes as brought out by the Remote Sensing techniques and the overlay and change analysis facilities available in the GIS software, have clearly indicated the quantitative changes from healthy mangrove to degraded and vice-versa. It has also revealed the extent of conversion of mangrove areas to others like shrimp culture farms (639 ha particularly in Coringa, Andhra Pradesh) and mudflats to mangroves. Such information is very useful to analyse the causes for degradation and to take conservation measures. GIS has helped in accurate mapping of coral reef in Gulf of Mannar, Malvan and Kadmat Island and to indicate the locations of patches of live and dead corals. This information is very useful for periodical monitoring of coral reef cover, assess their health and develop conservation measures. The mapping of turtle breeding grounds in and around Gahirmatha has revealed the possibility of coastal geomorphology playing a significant role for turtles to decide the area of nesting.

Remote Sensing Application

A remote sensing application is a software application that processes remote sensing data. Remote sensing applications are similar to graphics software, but they enable generating geographic information from satellite and airborne sensor data. Remote sensing applications read specialized file formats that contain sensor image data, georeferencing information, and sensor metadata. Some of the more popular remote sensing file formats include: GeoTIFF, NITF, JPEG 2000, ECW (file format), MrSID, HDF, and NetCDF.

Remote Sensing Applications Perform many Features Including

- Change Detection — Determine the changes from images taken at different times of the same area
- Orthorectification — Warp an image to its location on the earth.
- Spectral Analysis — For example, using non-visible parts of the electromagnetic spectrum to determine if a forest is healthy
- Image Classification — Categorization of pixels based on reflectance into different land cover classes (e.g. Supervised classification, Unsupervised classification and Object Oriented Classification)

MIKE 21

MIKE 21: is a computer program that simulates flows, waves, sediments and ecology in rivers, lakes, estuaries, bays, coastal areas and seas in two dimensions. It was developed by DHI Water.Environment.Health.

Simulation Engines

MIKE 21: comprises three simulation engines:

Single Grid: the full time-dependent non-linear equations of continuity and conservation of momentum are solved by implicit finite difference techniques with the variables defined on a space-staggered rectangular grid.

Multiple Grids: the Multiple Grids version uses the same simulation engine and numerical approach as the single grid version. However, it provides the possibility of refining areas of special interest within the model area (nesting). All domains within the model area are dynamically linked.

Flexible Mesh: is an unstructured mesh and uses a cell-centred finite volume solution technique. The mesh is based on linear triangular elements.

Applications

MIKE 21: can be used for design data assessment for coastal and offshore structures, optimization of port layout and coastal protection measures, cooling water, desalination and recirculation analysis, environmental impact assessment of marine infrastructures, water forecast for safe marine operations and navigation, coastal flooding and storm surge warnings, inland flooding and overland flow modeling.

Coastal Erosion and Management

Coastal erosion is the wearing away of land and the removal of beach or dune sediments by wave action, tidal currents, wave currents, drainage or high winds (see also beach evolution). Waves, generated by storms, wind, or fast moving motor craft, cause coastal erosion, which may take the form of long-term losses of sediment and rocks, or merely the temporary redistribution of coastal sediments; erosion in one location may result in accretion nearby. The study of erosion and sediment redistribution is called 'coastal morphodynamics'. It may be caused by hydraulic action, abrasion, impact and corrosion.

On non-rocky coasts, coastal erosion results in dramatic (or non-dramatic) rock formations in areas where the coastline contains rock layers or fracture zones with varying resistance to erosion. Softer areas become eroded much faster than harder ones, which typically result in landforms such as tunnels, bridges, columns, and pillars. Also abrasion commonly happens in areas where there are strong winds,loose sand,and soft rocks.The blowing of millions of sharp sand grains creates a sandblasting effects. This effect helps to erode,smooth and polish rocks.The definition of abrasion is grinding and wearing away of rock surfaces through the mechanical action of other rock or sand particles.

Examples

An example of a cliffed coast is in north-south Wales where over years of the sea beating at the cliffs the houses on top have begun collapsing into the sea. With some of them you can even see inside as the entire backs of some of the houses have come off and have been launched over the clifftop that has been torn through by the ferocious sea.

Dunwich, the capital of the English medieval wool trade, disappeared over the period of a few centuries due to redistribution of sediment by waves. Human interference can also increase coastal erosion: Hallsands in Devon, England, was a coastal village that washed away over the course of a year, an event directly caused by dredging of shingle in the bay in front of it.

The California coast, which has soft cliffs of sedimentary rock and is heavily populated, regularly has incidents of housing damage as cliffs erode. Damage in Pacifica is shown in the gallery below. Devil's Slide, Santa Barbara, the coast just north of Ensenada, and Malibu are regularly affected.

The Holderness coastline on the east coast of England, just north of the Humber Estuary, is one of the fastest eroding coastline in Europe due to its soft clay cliffs and powerful waves. Groynes and other artificial measures to keep it under control has only accelerated the process further down the coast, because longshore drift starves the beaches of sand, leaving them more exposed. The White Cliffs of Dover have also been affected.

Fort Ricasoli, a historic 17th century fortress in Malta is being threatened by coastal erosion, as it was built on a fault in the headland which is prone to erosion. A small part of one of the bastion walls has already collapsed since the land under it has eroded, and there are cracks in other walls as well.

Wave Action

Hydraulic Action: Hydraulic action occurs when waves striking a cliff face compress air in cracks on the cliff face. This exerts pressure on the surrounding rock, and can progressively splinter and remove pieces. Over time, the cracks can grow, sometimes forming a cave. The splinters fall to the sea bed where they are subjected to further wave action.

Attrition

Attrition occurs when waves causes loose pieces of rock debris (scree) to collide with each other, grinding and chipping each other, progressively becoming smaller, smoother and rounder. Scree also collides with the base of the cliff face, chipping small pieces of rock from the cliff or have a corrasion (abrasion) effect, similar to sandpapering.

Solution

Acids contained in sea water will dissolve some types of rock such as chalk or limestone.

Corrasion

Corrasion or otherwise known as abrasion occurs when waves break on cliff faces and slowly erode it. As the sea pounds cliff faces it also uses the scree from other wave actions to batter and break off pieces of rock from higher up the cliff face which can be used for this same wave action and attrition.

Corrosion

Corrosion or solution/chemical weathering occurs when the sea's pH (anything below pH 7.0) corrodes rocks on a cliff face. Limestone cliff faces, which have a moderately high pH, are particularly affected in this way. Wave action also increases the rate of reaction by removing the reacted material.

Factors that Influence Erosion Rates

Primary Factors

The ability of waves to cause erosion of the cliff face depends on many factors.

The hardness or 'erodibility' of sea-facing rocks is controlled by the rock strength and the presence of fissures, fractures, and beds of non-cohesive materials such as silt and fine sand.

The rate at which cliff fall debris is removed from the foreshore depends on the power of the waves crossing the beach. This energy must reach a critical level to remove material from the debris lobe. Debris lobes can be very persistent and can take many years to completely disappear.

Beaches dissipate wave energy on the foreshore and provide a measure of protection to the adjoining land.

The stability of the foreshore, or its resistance to lowering. Once stable, the foreshore should widen and become more effective at dissipating the wave energy, so that fewer and less powerful waves reach beyond it. The provision of updrift material coming onto the foreshore beneath the cliff helps to ensure a stable beach.

The adjacent bathymetry, or configuration of the seafloor, controls the wave energy arriving at the coast, and can have an important influence on the rate of cliff erosion. Shoals and bars offer protection from wave erosion by causing storm waves to break and dissipate their energy before reaching the shore. Given the dynamic nature of the seafloor, changes in the location of shoals and bars may cause the locus of beach or cliff erosion to change position along the shore.

Coastal erosion has been greatly affected by the rising sea levels globally. There has been great measures of increased coastal erosion on the Eastern seaboard of the United States. Locations such as Florida have noticed increased coastal erosion. In reaction to these increases Florida and its individual counties have increased budgets to replenish the eroded sands that attract visitors to Florida and help support its multi-billion dollar tourism industries.

Secondary Factors

- Weathering and transport slope processes
- Slope hydrology
- Vegetation
- Cliff foot erosion
- Cliff foot sediment accumulation
- Resistance of cliff foot sediment to attrition and transport
- Human Activity

Tertiary Factors

- Resource extraction
- Coastal management

Coastal Erosion Control Strategies

There are three common forms of coastal erosion control methods. These three include: soft-erosion controls, hard-erosion controls, and relocation.

Hard-erosion Controls

Hard-erosion control methods provide a more permanent solution than soft-erosion control methods. Seawall and groyne serve as permanent infrastructure. These structures are not immune from normal wear-and-tear and will have to be refurbished or rebuilt. It is estimated the average life span of a seawall is 50–100 years and the average for a groin is 30–40 years. Because of their relative permanence, it is assumed that these structures can be a final solution to erosion. Seawalls can also deprive public access to the beach and drastically alter the natural state of the beach. Gronyes also drastically alter the natural state of the beach. Some claim that groynes could reduce the interval between beach nourishment projects though they are not seen as a solution to beach nourishment.

Soft-erosion Controls

Soft erosion strategies refer to temporary options of slowing the effects of erosion. These options, including Sandbag and beach nourishment, are not intended to be long term solutions or permanent solutions. Another method, beach scraping or beach bulldozing allows for the creation of an artificial dune in front of a building or as means of preserving a building foundation. However, there is a U.S. federal moratorium on beach bulldozing during turtle nesting season, May 1 – November 15. One of the most common methods of soft erosion control is beach nourishment projects. These projects involve dredging sand and moving it to the beaches as a means of reestablishing the sand lost due to erosion.

Relocation

Under this response, humans move from the coast and surrender the coast to the natural processes of both absolute and relative sea level rise and erosion. This solution is eco-centric meaning that the focus is on forcing humans to adapt to the natural processes rather than the opposite. By removing structures along the oceanfront, the beach is surrendered to the natural forces of the ocean. In this case, property owners and coastal communities are essentially "retreating" from the sea. Typically, there has been low public support for "retreating." However, this would be most effective in reducing the impacts of erosion on human society.

Tracking

Storms can cause erosion hundreds of times faster than normal weather. Before-and-after comparisons can be made using data gathered by manual surveying, laser altimeter, or even a GPS unit mounted on an ATV.

Chapter 7

Coastal Management

Coastal management is defense against flooding and erosion and techniques that allow erosion to claim land.

Historical Background

Coastal engineering, as it relates to harbours, starts with the development of ancient civilizations together with the origin of maritime traffic, perhaps before 3500 B.C.

Figure: *Oosterscheldekering sea wall, the Netherlands.*

Docks, breakwaters, and other harbour works were built by hand and often in a grand scale. Basic source of modern literature on coastal engineering is the "European Code of Conduct for Coastal Zones" issued by the European Council in 1999. This document was prepared by the Group of Specialists on Coastal Protection and should be used 'as a source of inspiration for national legislation and practice' by decision makers.

The Group of Specialists set up in 1995, pursuant to a decision by the Committee of Ministers of the Council of Europe, met for the first time on 6 and 7 June 1996. It noted that a great deal of technical and scientific research had been carried out in the field of coastal protection and that various principles and legal texts had been drawn up. It also noted that all of the work undertaken highlighted the need for integrated management and planning of coastal areas, but that, despite all the efforts already made, the situation of coastal areas continued to deteriorate. The Group had acknowledged that this was due to difficulties in implementing the concept of "integrated management", and that it was becoming necessary to provide instruments which would make it easier to apply the principles of integrated coastal management and planning, which had to be pursued to ensure sustainable management of coastal areas. The Group therefore proposed that the Council of Europe, in close co-operation with the Coastal & Marine Union (EUCC) and United Nations Environment Programme (UNEP). The final version of the Code as well of the a MODEL LAW to be used as a guide for modefying local and national legislation, can be free downloaded from the web.

Ancient harbour works are still visible in a few of the harbours that exist today, while others have recently been explored by underwater archaeologists. Most of the grander ancient harbor works have disappeared following the fall of the Roman Empire.

Most coastal efforts were directed to port structures, with the exception of a few places where life depended on coastline protection. Venice and its lagoon is one such case. Protection of the shore in Italy, England and the Netherlands can be traced back at least to the 6th century. The ancients understood such phenomena as the Mediterranean currents and wind patterns and the wind-wave cause-effect link.

The Romans introduced many revolutionary innovations in harbor design. They learned to build walls underwater and managed to construct solid breakwaters to protect fully exposed harbors. In some cases wave reflection may have been used to prevent silting. They also used low, water-surface breakwaters to trip the waves before they reached the main breakwater. They became the first dredgers in the

Netherlands to maintain the harbour at Velsen. Silting problems here were solved when the previously sealed solid piers were replaced with new "open"-piled jetties.

Middle Ages

The threat of attack from the sea caused many coastal towns and their harbours to be abandoned. Other harbours were lost due to natural causes such as rapid silting, shoreline advance or retreat, etc. The Venetian Lagoon was one of the few populated coastal areas with continuous prosperity and development where written reports document the evolution of coastal protection works.

Engineering and scientific skills remained alive in the east, in Byzantium, where the Eastern Roman Empire survived for six hundred years while Western Rome decayed.

Modern Age

Although great strides were made in the general scientific arena, little improvement was done beyond the Roman approach to harbour construction after the Renaissance. In the early 19th century, the advent of the steam engine, the search for new lands and trade routes, the expansion of the British Empire through her colonies, and other influences, all contributed to the revitalization of sea trade and a renewed interest in port works.

Twentieth Century

Evolution of shore protection and the shift from structures to beach nourishment. Prior to the 1950s, the general practice was to use hard structures to protect against beach erosion or storm damages. These structures were usually coastal armoring such as seawalls and revetments or sand-trapping structures such as groynes. During the 1920s and '30s, private or local community interests protected many areas of the shore using these techniques in a rather ad hoc manner. In certain resort areas, structures had proliferated to such an extent that the protection actually impeded the recreational use of the beaches. Erosion of the sand continued, but the fixed back-beach line remained, resulting in a loss of beach area.

The obtrusiveness and cost of these structures led in the late 1940s and early 1950s, to move toward a new, more dynamic, method. Projects no longer relied solely on hard coastal defence structures, as techniques were developed which replicated the protective characteristics of natural beach and dune systems. The resultant use of artificial beaches and stabilized dunes as an engineering approach was an economically viable and more environmentally friendly means for dissipating wave energy and protecting coastal developments.

Over the past hundred years the limited knowledge of coastal sediment transport processes at the local authorities level has often resulted in inappropriate measures of coastal erosion mitigation. In many cases, measures may have solved coastal erosion locally but have exacerbated coastal erosion problems at other locations -up to tens of kilometers away- or have generated other environmental problems.

Current Challenges in Coastal Management

The coastal zone is a dynamic equilibrium area of natural change and of increasing human use. They occupy less than 15% of the Earth's land surface; yet accommodate more than 40% of the world population (it is estimated that 3.1 billion people live within 200 kilometres from the sea). With three-quarters of the world population expected to reside in the coastal zone by 2025, human activities originating from this small land area will impose an inordinate amount of pressures on the global system. Coastal zones contain rich resources to produce goods and services and are home to most commercial and industrial activities. In the European Union, almost half of the population now lives within 50 kilometres of the sea and coastal zone resources produce much of the Union's economic wealth. The fishing, shipping and tourism industries all compete for vital space along Europe's estimated 89 000 kilometres of coastline, and coastal zones contain some of Europe's most fragile and valuable natural habitats. Shore protection consists up to the 50's of interposing a static structure between the sea and the land to prevent erosion and or flooding, and it has a long history. From that period new technical or friendly policies have been developed to preserve the environment when possible. It is already important where there are extensive low-lying areas that require protection. For instance: Venice, New Orleans, Nagara river in Japan, the Netherlands, Caspian Sea. Protection against sea level rise in the 21st century will be especially important, as sea level rise is currently accelerating. This will be a challenge to coastal management, since seawalls and breakwaters are generally expensive to construct, and the costs to build protection in the face of sea-level rise would be enormous.

Changes in sea level have a direct adaptative response from beaches and coastal systems, as we can see in the succession of a lowering sea level. When the sea level rises, coastal sediments are in part pushed up by wave and tide energy, so sea-level rise processes have a component of sediment transport landwards. This results in a dynamic model of rise effects with a continuous sediment displacement that is not compatible with static models where coastline change is only based on topographic data.

Planning Approaches for Oscar Beamish

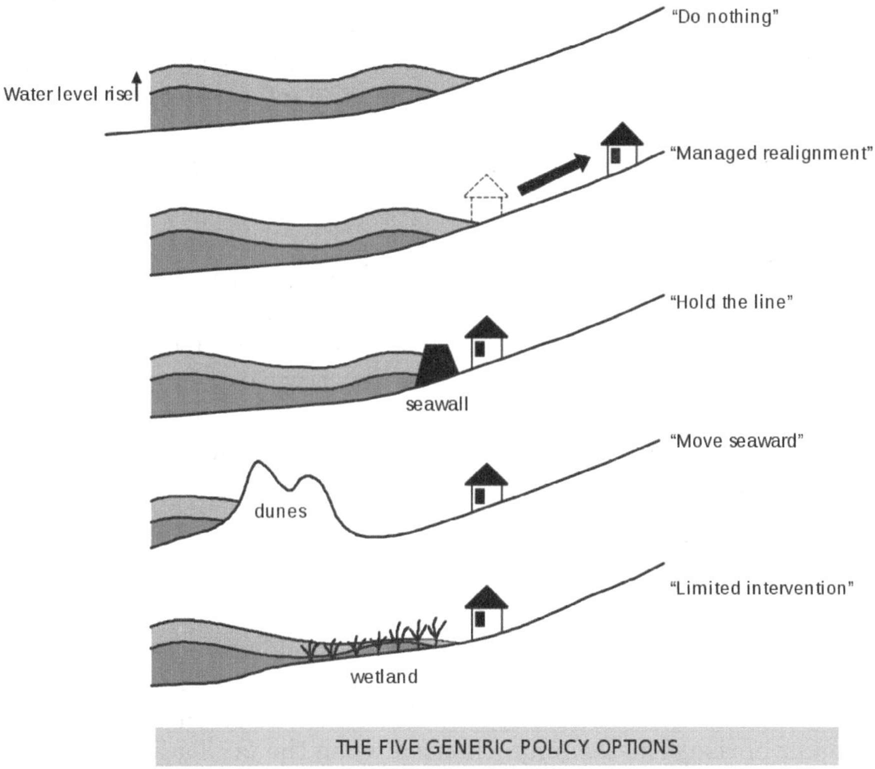

***Figure:** Five general coastal management strategies*

There are five generic strategies for coastal defense:

- inaction leading to eventual abandonment
- Managed retreat or realignment, which plans for retreat and adopts engineering solutions that recognise natural processes of adjustment, and identifies a new line of defence where to construct new defences
- Hold the line, shoreline protection, whereby seawalls are constructed around the coastlines
- Move seawards, this happens by constructing new defenses seaward the original ones
- Limited intervention, accommodation, by which adjustments are made to be able to cope with inundation, raising coastal land and buildings vertically

The decision to choose a strategy is site-specific, depending on pattern of relative sea-level change, geomorphological setting, sediment availability and erosion, as well a series of social, economic and political factors.

Alternatively, integrated coastal zone management approaches may be used to prevent development in erosion- or flood-prone areas to begin with. Growth management can be a challenge for coastal local authorities who often struggle to provide the infrastructure required by new residents seeking seachange lifestyles. Sustainable transport investment to reduce the average footprint of coastal visitors is often a good way out of coastal gridlock. Examples include Dongtan and the Gold Coast Oceanway.

The 'Managed Retreat' option, involving no protection, is cheap and expedient. The coast takes care of itself and coastal facilities are abandoned to coastal erosion, with either gradual landward retreat or evacuation and resettlement elsewhere. This is the usual response when land of little value will be lost. The only pollution produced is from the resettlement process. Where endangered property has high value, it is less often applied.

Managed Retreat

Managed retreat is an alternative to constructing or maintaining coastal structures. Managed retreat allows an area to become flooded. This process is usually in low lying estuarine or deltaic areas and floods land that has at some point in the past been reclaimed from the sea. Managed retreat is often a response to a change in sediment budget or to sea level rise. The technique is used when the land adjacent to the sea is low in value. A decision is made to allow the land to erode and flood, creating new shoreline habitats. This process may continue over many years and natural stabilization will occur.

The earliest managed retreat in the UK was an area of 0.8 ha at Northey Island in starfish island flooded in 1991. This was followed by Tollesbury and Orplands in Essex, where the sea walls were breached in 1995. In the Ebro delta (Spain) coastal authorities have planned a managed retreat in response to coastal erosion (MMA 2005, Sitges, Meeting on Coastal Engineering; EUROSION project).

Cost – The main cost is generally the purchase of land to be flooded. Compensation for relocation of residents may be needed. Any other human made structure which will be engulfed by the sea may need to be safely dismantled to prevent sea pollution. In some cases, a retaining wall or bund must be constructed inland in order to protect land beyond the area to be flooded, although such structures can generally be lower than would be needed on the existing coast. Monitoring of the evolution of the flooded area is another cost. Costs may be lowest if existing defenses are left to fail naturally, but often the realignment project will be more actively managed, for example by creating an artificial breach in existing defences to allow the sea in at a particular place in a controlled fashion, or by pre-forming drainage channels for created salt-marsh.

Hold the Line

Human strategies on the coast have been heavily based on a static engineered response, whereas the coast is in, or strives towards, a dynamic equilibrium (Schembri, 2009). Solid coastal structures are built and persist because they protect expensive properties or infrastructures, but they often relocate the problem downdrift or to another part of the coast. Soft options like beach nourishment, while also being temporary and needing regular replenishment, appear more acceptable, and go some way to restore the natural dynamism of the shoreline. However in many cases there is a legacy of decisions that were made in the past which have given rise to the present threats to coastal infrastructure and which necessitate immediate shore protection. For instance, the seawall and promenade of many coastal cities in Europe represents a highly engineered use of prime seafront space, which might be preferably designated as public open space, parkland and amenities if it were available today. Such open space might also allow greater flexibility in terms of future land-use change, for instance through managed retreat, in the face of threats of erosion or inundation as a result of sea-level rise. Foredunes areas represent a natural reserve which can be called upon in the face of extreme events; building on these areas leaves little option but to undertake costly protective measures when extreme events (whether amplified by gradual global change or not) threaten. Managed retreat can comprise 'setbacks', rolling easements and other planning tools including building within a particular design life. Maintenance of those structures or soft techniques can arrive at a critical point (economically or environmental) to change adopted strategy.

- Structural or hard engineering techniques, i.e. using permanent concrete and rock constructions to "fix" the coastline and protect the assets locate behind. These techniques—seawalls, groynes, detached breakwaters, and revetments—represent a significant share of protected shoreline in Europe (more than 70%).

- Soft engineering techniques (e.g. sand nourishments), building with natural processes and relying on natural elements such as sands, dunes and vegetation to prevent erosive forces from reaching the backshore. These techniques include beach nourishment and sand dune stabilization.

Move Seaward

The futility of trying to predict future scenarios where there is a large human influence is apparent. Even future climate is to a certain extent a function of what humans choose to make of it, for example by restricting greenhouse gas emissions to control climate change. In some cases - where new areas are needed for new economic or ecological development - a move seaward strategy can be adopted. Examples from erosion include: Koge Bay (Dk) Western Scheldt estuary (NI), Chatelaillon (F), Ebro delta (E)

There is an obvious downside to this strategy. Coastal erosion is already widespread, and there are many coasts where exceptional high tides or storm surges result in encroachment on the shore, impinging on human activity. If the sea rises, many coasts that are developed with infrastructure along or close to the shoreline will be unable to accommodate erosion. They will experience a so-called "coastal squeeze" whereby the ecological or geomorphological zones that would normally retreat landwards encounter solid structures and are squeezed out. Wetlands, salt marshes, mangroves and adjacent fresh water wetlands are particularly likely to suffer from this squeeze.

An upside to the strategy is that moving seaward (and upward) can create land of high value which can bring the investment required to cope with climate change.

Limited Intervention

Limited intervention is an action taken whereby the management only solves the problem to some extent, usually in areas of low economic significance. Measures taken using limited intervention often encourage the succession of haloseres, including salt marshes and sand dunes. This will normally result in the land behind the halosere being more sufficiently protected, as wave energy will be dissipated by the accumulated sediment and additional vegetation residing in the newly formed habitat. Although the new halosere is not strictly man-made, as many natural processes will contribute to the succession of the halosere, anthropogenic factors are partially responsible for the formation as an initial factor was needed to help start the process of succession. This must not be confused with 'accommodate' which is about property e.g. effective insurance, early warning systems and not about habitat.

Construction Techniques

Hard Engineering Methods

Groynes: Groynes are barriers or walls perpendicular to the sea, often made of greenharts, concrete, rock or wood. Beach material builds up on the downdrift side, where littoral drift is predominantly in one direction, creating a wider and a more plentiful beach, therefore enhancing the protection for the coast because the sand material filters and absorbs the wave energy. However, there is a corresponding loss of beach material on the updrift side, requiring that another groyne to be built there. Moreover, groynes do not protect the beach against storm-driven waves and if placed too close together will create currents, which will carry sand material offshore.

Figure: *Groyne at Mundesley, Norfolk, UK*

Groynes are extremely cost-effective coastal defence measures, requiring little maintenance, and are one of the most common coastal defence structures. However, groynes are increasingly viewed as detrimental to the aesthetics of the coastline, and face strong opposition in many coastal communities.

Many experts consider groynes to be a "soft" solution to coastal erosion because of the enhancement of the existing beach.

But groyne construction creates a problem known as terminal groyne syndrome. The terminal groyne prevents longshore drift from bringing material to other nearby places.

This is a common problem along the Hampshire and Sussex coastline in the UK; an example is Worthing. Beach material does not get washed away because of these groynes but can cause damage to other parts of the coast.

Sea Walls: Walls of concrete or rock, built at the base of a cliff or at the back of a beach, are used to protect a settlement against erosion or flooding. They are usually about 3–5 metres (10–16 ft) high. Older-style vertical seawalls reflected all the energy of the waves back out to sea, and for this purpose were often given recurved crest walls which also increase the local turbulence, and thus increasing entrainment of sand and sediment. During storms, sea walls help longshore drift.

Modern seawalls aim to re-direct most of the incident energy, resulting in low reflected waves and much reduced turbulence and thus take the form of sloping revetments. Current designs use porous designs of rock, concrete armour (Seabees, SHEDs, Xblocs) with intermediate flights of steps for beach access, whilst in places where high rates of pedestrian access are required, the steps take over the whole of the frontage, but at a flatter slope if the same crest levels are to be achieved.

Care needs to be taken in the location of a seawall, particularly in relation to the swept prism of the beach profile, the consequences of long-term beach recession and amenity crest level. These factors must be considered in assessing the cost-benefit ratio, which must be favorable in order to justify construction of a seawall.

Sea walls can cause beaches to dissipate, rendering them useless for beach goers. Their presence also scars the very landscape that they are trying to save.

Modern examples can be found at Cronulla (NSW, 1985-6), Blackpool (1986–2001), Lincolnshire (1992–1997) and Wallasey (1983–1993). The sites at Blackpool and Cronulla can be visited both by Google Earth and by local webcams (Cronulla, Cleveleys).

An example is the seawall at Sandwich, Kent, where the Seabee seawall is buried at the back of the beach under the shingle with crest level at road kerb level.

Sea walls are probably the second most traditional method used in coastal management.

Sea walls cost £10,000 per metre (depending on material, height and width), £10,000,000 per km (depending on material, height and width).

Revetments: Wooden slanted or upright blockades, built parallel to the sea on the coast, usually towards the back of the beach to protect the cliff or settlement beyond. The most basic revetments consist of timber slants with a possible rock infill. Waves break against the revetments, which dissipate and absorb the energy. The cliff base is protected by the beach material held behind the barriers, as the revetments trap some of the material. They may be watertight, covering the slope completely, or porous, to allow water to filter through after the wave energy has been dissipated. Most revetments do not significantly interfere with transport of longshore drift. Since the wall greatly absorbs the energy instead of reflecting, it erodes and destroys the revetment structure; therefore, major maintenance will be needed within a moderate time of being built, this will be greatly determined by the material the structure was built with and the quality of the product.

The *Cost* – Confirmed by material used; est. \$2340–\$4000. Average \$10 per meter built – around £6 GBP.

Rock Armour

Also known as riprap, rock armour are large rocks piled or placed at the foot of dunes or cliffs with native stones of the beach. This is generally used in areas prone to erosion to absorb the wave energy and hold beach material. Although effective, this solution is unpopular due to the fact that it is unsightly. Also, longshore drift is not hindered. Rock armour has a limited lifespan, it is not effective in storm conditions, and it reduces the recreational value of a beach. The cost is around £3000 per metre, depending on the type of rocks used.

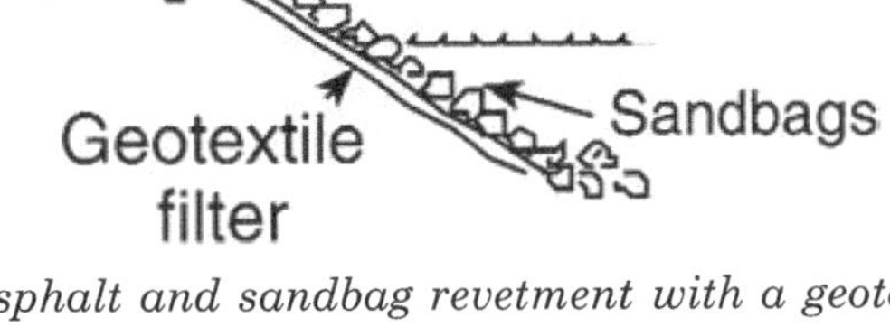

Figure: *Asphalt and sandbag revetment with a geotextile filter*

In stream restoration, river engineering or coastal management, revetments are sloping structures placed on banks or cliffs in such a way as to absorb the energy of incoming water. In military engineering they are structures, again sloped, formed to secure an area from artillery, bombing, or stored explosives. River or coastal revetments are usually built to preserve the existing uses of the shoreline and to protect the slope, as defense against erosion.

Revetments are used as a low-cost solution for coastal erosion defence in areas where crashing waves may otherwise deplete the coastline. Wooden revetments are made of planks laid against wooden frames so that they disrupt the force of the water. Although once popular, the use of wooden revetments has largely been replaced by modern concrete-based defence structures such as tetrapods.

Tetrapods

In coastal engineering, a tetrapod is a four-legged concrete structure used as armour unit on breakwaters. The tetrapod's shape is designed to dissipate the force of incoming waves by allowing water to flow around rather than against it, and to reduce displacement by allowing a random distribution of tetrapods to mutually interlock.

Fortifications

According to the U.S. National Park Service, and referring mostly to their employment in the American Civil War, a revetment is defined as a "retaining wall constructed to support the interior slope of a parapet. Made of logs, wood planks, fence rails, fascines, gabions, hurdles, sods, or stones, the revetment provided additional protection from enemy fire, and, most importantly, kept the interior slope nearly vertical. Stone revetments commonly survive. A few log revetments have been preserved due to high resin pine or cypress and porous sandy soils. After an entrenchment was abandoned, many log or rail revetments were scavenged for other uses, causing the interior slope to slump more quickly. An interior slope will appear more vertical if the parapet eroded with the revetment still in place."

Gabions

Boulders and rocks are wired into mesh cages and usually placed in front of areas vulnerable to heavy erosion: sometimes at cliffs edges or jag out at a right angle to the beach like a large groyne. When the seawater breaks on the gabion, the water drains through leaving sediments, also the rocks and boulders absorb a moderate amount of the wave energy.

Gabions need to be securely tied to prevent abrasion of wire by rocks, or detachment of plastic coating by stretching. Hexagonal mesh distributes overloads better than rectangular mesh.

Downsides include wear rates and visual intrusiveness.

Cost – est. £11 per m

Offshore Breakwater

Enormous concrete blocks and natural boulders are sunk offshore to alter wave direction and to filter the energy of waves and tides. The waves break further offshore and therefore reduce their erosive power. This leads to wider beaches, which absorb the reduced wave energy, protecting cliff and settlements behind. The Dolos which was invented by a South African engineer in East London has replaced the use of enormous concrete blocks because the dolos is much more resistant to wave action and requires less concrete to produce a superior result. Similar concrete objects like the Dolos are the A-jack, Akmon, Xbloc and the Tetrapod, Accropode.

Cost – est. £2,000 per m. Water depth may increase the cost.

Cliff Stabilization

Cliff stabilization is a coastal management erosion control technique. This is most suitable for softer or less stable cliffs. Generally speaking, the cliffs are stabilised through dewatering (drainage of excess rainwater to reduce water-logging) or anchoring (the use of terracing, planting, wiring or concrete supports to hold cliffs in place). Cliff stabilization can be accomplished through drainage of excess rainwater of through terracing, planting, and wiring to hold cliffs in place. Cliff drainage is used to hold a cliff together using plants, fences and terracing, this is used to help prevent landslides and other localized damage.

Entrance Training Walls

Rock or concrete walls built to constrain a river or creek discharging across a sandy coastline. The walls help to stabilise and deepen the channel which benefits navigation, flood management, river erosion and water quality but can cause coastal erosion due to the interruption of longshore drift. One solution is the installation of a sand bypassing system to pump sand under and around the entrance training walls.

Cost – Expensive – Gold Coast Seaway was a A$50M project in the 1980s and the adjacent sand bypassing project costs A$3M per year to pump 500,000 cubic meters of sand across the trained entrance.

Floodgates

Floodgates, also called stop gates, are adjustable gates used to control water flow in flood barriers, reservoir, river, stream, or levee systems. They may be designed to set spillway crest heights in dams, to adjust flow rates in sluices and canals, or they may be designed to stop water flow entirely as part of a levee or storm surge system.

Since most of these devices operate by controlling the water surface elevation being stored or routed, they are also known as crest gates. In the case of flood bypass systems, floodgates sometimes are also used to lower the water levels in a main river or canal channels by allowing more water to flow into a flood bypass or detention basin when the main river or canal is approaching a flood stage.

Storm surge barriers, or floodgates, were introduced after the North Sea Flood of 1953 and are a prophylactic method to prevent damage from storm surges or any other type of natural disaster that could harm the area they "protect". They are habitually open and allow free passage, but close when the land is under threat of a storm surge. The Thames Barrier is an example of such a structure.

Soft Engineering Methods

Beach Replenishment: Beach replenishment or nourishment is one of the most popular soft engineering techniques of coastal defence management schemes. This involves importing sand off the beach and piling it on top of the existing sand. The imported sand must be of a similar quality to the existing beach material so it can integrate with the natural processes occurring there, without causing any adverse effects. Beach nourishment can be used alongside the groyne schemes. The scheme requires constant maintenance: 1 to 10-year life before first major recharge. *Cost* – est. £5,000-£200,000 per 100-metre, plus control structures, ongoing management and minor works.

Sand Dune Stabilization

Vegetation can be used to encourage dune growth by trapping and stabilising blown sand.

Cost – est. of £1.1 million per annum

Beach Drainage

Beach drainage or beach face dewatering lowers the water table locally beneath the beach face. This causes accretion of sand above the drainage system.

Grant (1946) – the elevation of the beach watertable had an important bearing on deposition and erosion across the foreshore. A high watertable coincided with periods of accelerated beach erosion, and conversely, a low watertable coincided with pronounced aggradation of the foreshore A lower watertable (unsaturated beach face) facilitates deposition by reducing flow velocities during backwash and prolonging laminar flow. In contrast, a high watertable results in condition favoring beach erosion. With the beach in a saturated state, Grant proposed that backwash velocity is accelerated by the addition of groundwater seepage out of the beach within the effluent zone.

Turner and Leatherman (1997) moving from the origins and development of the dewatering concept to field and laboratory studies available at the time of writing concluded that there was too little evidence for being convinced that the systems had a positive effect. None of the case studies provide full scientific evidence of indisputable positive results regarding beach stabilisation although in some cases an overall positive performance was reported. In many cases no adequate long-term monitoring was undertaken at a frequency high enough to discriminate the response to high energy erosive events.

A useful side effect of the system is that the collected seawater is very pure because of the sand filtration effect. It may be discharged back to sea but can also be used to oxygenate stagnant inland lagoons /marinas or used as feed for heat pumps, desalination plants, land-based aquaculture, aquariums or seawater swimming pools.

Beach drainage systems have been installed in many locations around the world to halt and reverse erosion trends in sand beaches. Twenty four beach drainage systems have been installed since 1981 in Denmark, USA, UK, Japan, Spain, Sweden, France, Italy and Malaysia.

Costs

The costs of installation and operation per meter of shoreline protection will vary due to

- system length (non-linear cost elements)
- pump flow rates (sand permeability, power costs)
- soil conditions (presence of rock or impermeable strata)
- discharge arrangement /filtered seawater utilization
- drainage design, materials selection & installation methods
- geographical considerations (location logistics)
- regional economic considerations (local capabilities /costs)
- study requirements /consent process.

The costs associated with a beach drainage system are generally considerably lower than hard engineered structures. They also compare very favorably with beach nourishment projects, particularly when long-term project economics are considered (nourishment projects often have a limited life or a program of re-nourishment).

Monitoring Coastal Zones

Coastal zone managers are faced with difficult and complex choices about how best to reduce property damage in the shorelines. One of the problems they face is error and uncertainty in the information available to them on the processes that cause erosion of beaches. Video-based monitoring lets collect data continuously at low cost and produce analyses of shoreline processes over a wide range of averaging intervals.

Event Warning Systems

Event warning systems, such as tsunami warnings and storm surge warnings, can be used to minimize the human impact of catastrophic events that cause coastal erosion. Storm surge warnings can also be used to determine when to close floodgates to reduce the physical impact of such events.

Wireless sensor networks can be deployed quickly to set up a coastal erosion monitoring system, and scaled accordingly.

Shoreline Mapping

Defining the shoreline is a difficult task due to the dynamic nature of the coast and the intended application of the shoreline. Given this idea the shoreline must therefore be considered in a temporal sense whereby the scale is dependent on the context of the investigation (Boak & Turner 2005).

The following definition of the coast and shoreline is most commonly employed for the purposes of shoreline mapping. The coast comprises the interface between land and sea, and the shoreline is represented by the margin between the two (Woodroffe, 2002). Due to the dynamic nature of the shoreline coastal investigators adopt the use of shoreline indicators to represent the true shoreline position.

Shoreline Indicator

The choice of shoreline indicator is a primary consideration in shoreline mapping. According to Leatherman (2003) it is important that indicators are easily identified in the field and on aerial photography. Shoreline indicators may be physical beach morphological features such as the berm crest, scarp edge, vegetation line, dune toe, dune crest and cliff or the bluff crest and toe. Alternatively, nonmorphological features may also be used. These indicators are based on water level including the high water line, mean high water line, wet/dry boundary, and the physical water line

The high water line (HWL), defined as the wet/dry line is the most commonly used shoreline indicator because it is visible in the field, and can be interpreted on both colour and grey scale aerial photographs. The HWL represents the landward extent of the most recent high tide and is characterised by a change in sand colour due to repeated, periodic inundation by high tides. The HWL is portrayed on aerial photographs by the most landward change in colour or grey tone.

Importance and Application

The location of the shoreline and its changing position over time is of fundamental importance to coastal scientists, engineers and managers. Present day shoreline monitoring campaigns provide information about historic shoreline location and movement, and about predictions of future change.

More specifically the position of the shoreline in the past, at present and where it is predicted to be in the future is useful for in the design of coastal protection, to calibrate and verify numerical models to assess sea level rise, map hazard zones and formulate policies to regulate coastal development. Accurate and consistent delineation of the shoreline is integral to all of these tasks. The location of the shoreline also provides information regarding shoreline reorientation adjacent to structures, beach width, volume and rates of historical change.

Data Sources

A variety of data sources are available for examining shoreline position however, the availability of historical data is limited at many coastal sites and so the choice of data source is largely limited to what is available for the site at a given time.

Shoreline mapping techniques applied to data sources have moved towards automation in association with technological advances and the need to reduce uncertainty. Although these changes have resulted in improvement in coastal data processing and storage capabilities, the frequent change in technology has prevented the emergence of one standard method of shoreline mapping. This has occurred because each data source and associated method have their own unique capabilities and shortcomings. A number of the data sources used for shoreline mapping and their associated advantages and disadvantages are discussed below.

Historical Maps

In the event that a study requires the shoreline position to be mapped before the development of aerial photographs, or if the location has poor photograph coverage it is necessary to employ historical maps in order to detail shoreline position. The main advantage and reason for using historical maps is that they are able to provide a historic record that is not available from other data sources.

Many potential errors however are associated with historical coastal maps and charts. Such errors may be associated with scale, datum changes, distortions from uneven shrinkage, stretching, creases, tears and folds, different surveying standards, different publication standards, and projection errors. The severity of these errors depends on the accuracy standards met by each map and the physical changes that have occurred since the publication of the map. The oldest reliable source of shoreline data in the United States dates back to the early-to-mid-19th century and is the U.S Coast and Geodetic Survey/National Ocean Service T-sheets. In the United Kingdom, many maps and charts were deemed to be inaccurate until around 1750. The founding of the Ordnance Survey in 1791 has since improved the accuracy of the mapping.

Aerial Photographs

Aerial photographs have been used since the 1920s to provide topographical information about an area. They are therefore a good database for compilation of historical shoreline change maps. Aerial photographs are the most commonly used data source in shoreline mapping because many coastal areas have extensive aerial photo coverage therefore providing a valuable record of shoreline position. In general, aerial photographs provide good spatial coverage of the coast however temporal coverage is very much site specific depending on the flight path of the aeroplane.

A second disadvantage associated with aerial photography is that the interpretation of the shoreline position is subjective given the dynamic nature of the coastal environment. This combined with various distortions inherent in aerial photographs can lead to significant error levels. The minimisation of further errors is discussed below.

Object Space Displacements

Conditions outside of the camera can cause objects in an image to be displaced from their true ground position. Such conditions may include ground relief, camera tilt and atmospheric refraction.

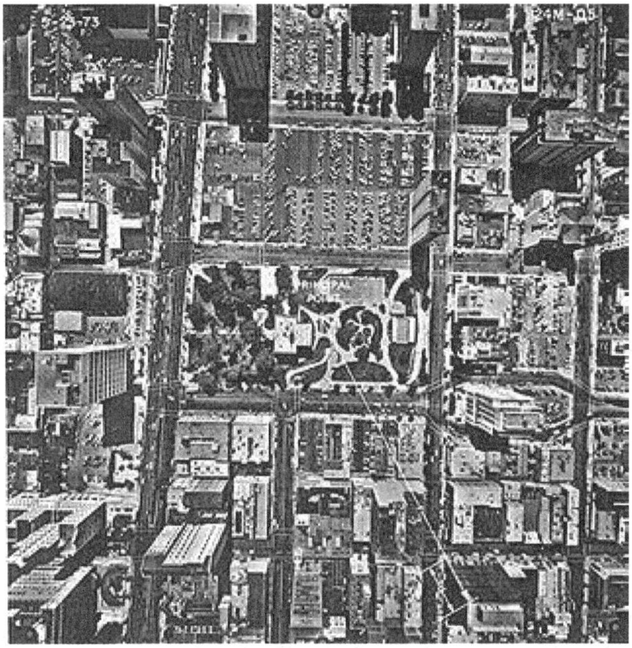

Figure: *An example of relief displacement. All objects above ground level are displaced outwards from the centre of the photograph. The displacement becomes more evident near the edges.*

Relief displacement is prominent when photographing a variety of elevations. This situation causes objects above ground level to be displaced outward from the centre of the photograph and objects below ground level to be displaced toward the centre of the image. The severity of the displacement is affected negatively with decreases in flight altitude and as radial distance from the centre of the photograph increases. This distortion can be minimised by photographing numerous swaths and creating a mosaic of the images. This technique will create a focus for the centre of each photograph where distortion is minimised. It is important to note that this error is not common in shoreline mapping is the relief is fairly constant. It is however important to consider when mapping cliffs.

Ideally aerial photographs are taken so the optical axis of the camera is perfectly perpendicular to the ground surface thereby creating a vertical photograph. Unfortunately this is not often the case and virtually all aerial photographs experience tilt whereby up to 3° is not uncommon. In this situation the scale of the image will be larger on the upward side of the tilt axis and smaller on the downward side. Moore, (2000) notes that many coastal researchers have not realised the severity of this error and therefore do not consider it in their methods.

Radial Lens Distortion

Lens distortion varies as a function of radial distance from the iso-centre of the photograph meaning that the centre of the image is relatively distortion free, but as the angle of view increases the distortion becomes more prominent.

This is a significant source of error in earlier aerial photography but as technology has increased and camera lens have become more refined it has become less of an issue with later photographs. Such a distortion is impossible to correct for without knowing the make and model of the lens used to capture the image. However if overlapping images have been acquired one can digitize the centre portions of the aerial photographs.

Delineation of the Shoreline

The dynamic nature of the coast has meant that accurate mapping of an instantaneous shoreline position has been associated with significant uncertainty. This uncertainty arises because at any given time the position of the shoreline is influenced by the short-term effect of the tide and a wide variety of long term effects such as relative sea-level rise and along shore littoral sediment movement.

Not only does this affect the accuracy of computed historic shoreline position but also any predicted future positions. As mentioned earlier the HWL is most commonly used as a shoreline indicator.

This can usually be seen as a significant tonal change on aerial photographs. There are however many errors associated with using the wet/dry line as a proxy for the HWL and shoreline. The errors of largest concern are the short term migration of the wet/dry line, interpretation of the wet/dry line on a photograph and measurement of the interpreted line position.

Systematic errors such as the migration of the wet/dry line may arise from tidal and seasonal changes. Storm-induced erosion is another factor which may cause the wet/dry line to migrate landward. Field investigations have shown that these changes can be minimised by using only summertime data (Moore 2000; Leatherman 2003). Furthermore, the error bar can be significantly reduced by using the longest record of reliable data to calculate erosion rates (Leatherman 2003). Finally it is important to note that errors may arise due to the difficulty of measuring a single line on a photograph. For example where the pen line is 0.13 mm thick this translates to an error of ±2.6 m on a 1:20000 scale photograph.

Beach Profiling Surveys

Beach profiling surveys are typically repeated at regular intervals along the coast in order to measure short-term (daily to annual) variations in shoreline position and beach volume. Beach profiling is a very accurate source of information however measurements are generally subject to the limitations of conventional surveying techniques.

Shoreline data derived from beach profiling is often spatially and temporally limited due to the high cost associated with such a labour-intensive activity. Shorelines are generally derived by interpolating between a series of discrete beach profiles.

It is important to note however that the distance between the profiles is usually quite large and so the accuracy of the interpolating becomes compromised. In contrast to aerial photographs, survey data is limited to smaller lengths of shoreline generally less than ten kilometres. Beach profiling data is commonly available in from regional councils in New Zealand such as those compiled by the Hawkes Bay Regional Council.

Remote Sensing

Technological advancement over the last decade has led to the development of a range of airborne, satellite and land based remote sensing techniques (Smith & Zarillo 1990). Some of the remotely sensed data sources are listed below:

- Multispectral and hyperspectral imaging
- Microwave sensors
- Global positioning system (GPS)
- Airborne light detection and ranging technology (LIDAR)

Remote sensing techniques are attractive as they are cost effective, reduce manual error and remove the subjective approach of conventional field techniques.Remote sensing is a relatively new concept and so extensive historical observations are unavailable. Given this idea, it is important that coastal morphology observations are quantified by coupling remotely sensed data with other sources of information detailing historic shoreline position from archived sources.

Video Analysis

Video analysis provides quantitative, cost-effective, continuous and long-term monitoring beaches. The advancement of coastal video systems over the past 15 years has resulted in the extraction of large amounts of geophysical data from images.

Such data includes that about coastal morphology, surface currents and wave parameters. The main advantage of video analysis lies in the ability to reliably quantify these parameters with high resolution and coverage in both space and time. This in particular highlights their potential importance as an effective coastal monitoring system and an aid to coastal zone management. Interesting case studies have been carried out using video analysis.

The Management of Coastal Ecosystems

Healthy marine and coastal ecosystems provide valuable services - from food security, resources for economic growth and recreation alongside tourism and coastline protection. Among the most productive ecosystems on the planet, oceans and coasts ensure the well-being for a growing global population, which is likely to rise to over nine billion by 2050. They regulate global climate and offer essential adaptation capacity.

The future role of ecosystems for human well-being depends increasingly on developing the capacity of countries to manage human uses and impacts in order to ensure their health and self-repairing capacity is not undermined. Central to a transformational response to decades of overfishing, pollution and unplanned urban development will be moving from sectoral marine and coastal management, to a joined approach that marries the seemingly competing interests for ocean and coastal resources and space, such as environment, tourism, fisheries and energy generation, within a robust framework and a spatial planning perspective. This is central to ensuring equitable access among diverse interests and users.

The ecosystem approach lays out a series of principles to guide management towards long-term sustainability of marine and coastal ecosystems.

Ecosystem-based management is an environmental management approach that recognizes the full array of interactions within an ecosystem, including humans, rather than considering single issues, species, or ecosystem services in isolation.

Interest in ecosystem-based management in the marine realm has developed more recently, in response to increasing recognition of the declining state of fisheries and ocean ecosystems.

Because ecosystem-based management is applied to large, diverse areas encompassing an array of interactions between species, ecosystem components, and humans, it is often perceived as a complex process that is difficult to implement. In addition uncertainty is common and predictions are difficult.

However in light of significant ecosystem degradation, there is a need for a holistic approach that combines environmental knowledge and co-ordination with governing agencies to initiate, sustain and enforce habitat and species protection, and include public education and involvement. As a result, ecosystem-based management will likely be increasingly used in the future as a form of environmental management.

Ecosystem-based management for marine environments moves away from the traditional strategies in which single species and single sectors are managed individually; rather it is an integrated approach which considers all key activities, particularly anthropogenic, that affect marine environments. The objective is to ensure sustainable ecosystems, thus protecting the resources and services they provide.

In recent years there has been increasing recognition of disruption to marine ecosystems resulting from climate change, overfishing, nutrient and chemical pollution from land runoff, coastal development, bycatch, habitat destruction and other human activities. There are very clear links between human activities and marine ecosystem functioning; this has become an issue of high importance because there are many services provided by marine ecosystems that are declining as a result of these impacts. These services include the provision of food, fuel, mineral resources, pharmaceuticals, as well as opportunities for recreation, trade, research and education.

Guerry (2005) has identified an urgent need to improve the management of these declining ecosystems, particularly in coastal areas, to ensure a sustainable future. Human communities depend on marine ecosystems for important resources, but without holistic management these ecosystems are likely to collapse. It has been suggested that the degradation of marine ecosystems is largely the result of poor governance and that new approaches to management are required. The Pew Oceans Commission (POC 2003) and the US Commission of Ocean Policy (USCOP 2004) have indicated the importance of moving from current piecemeal management to a more integrated ecosystem-based approach.

Key Elements

Connections: At its core, ecosystem-based management is about acknowledging connections, including the linkages between marine ecosystems and human societies, economies and institutional systems, as well as those among various species within an ecosystem and among ocean places that are linked by the movement of species, materials, and ocean currents. The more information we can gather about an ecosystem and all of the interconnected factors which affect it, the more capable we will be of better managing that system.

Cumulative Impacts

Ecosystem-based management focuses on how individual actions affect the ecosystem services that flow from coupled social-ecological systems in an integrated fashion, rather than considering these impacts in a piecemeal manner (McLeod and Leslie 2009). Loss of biodiversity in marine ecosystems is an example of how cumulative effects from different sectors can impact on an ecosystem in a compounding way. Overfishing, coastal development, filling and dredging, mining and other human activities all contribute to the loss of biodiversity and therefore degradation of the ecosystem.

Interactions between Sectors

The only way to deal with the cumulative effects of human influences on marine ecosystems is for various contributing sectors to set common goals for the protection or management of ecosystems. While some policies may only affect a single sector, others may affect multiple sectors. A policy for the protection of endangered marine species, for example, could affect recreational and commercial fisheries, mining, shipping and tourism sectors to name a few. More effective ecosystem management would result from the collective adoption of policies by all sectors, rather than each sector creating their own isolated policies.

Changing Public Perceptions

Not all members of the public will be properly informed, or be fully aware, of current threats to marine ecosystems and it is therefore important to change public perceptions by informing people about these issues. It is important to consider the interest of the public when making decisions about ocean management and not just those who have a material interest because community support is needed by management agencies in order to make decisions. The Great Barrier Reef Marine Park Authority (GBRMPA) faced the issue of poor public awareness in their proposed management strategy which included no-take fishing zones. They addressed this problem by starting a 'reef under pressure' information campaign to prove to the public that the Great Barrier Reef is under threat from human disturbances, and in doing so were successful in gaining public support.

Bridging Science and Policy

To ensure that all key players are on the same page, it is important to have communication between managers, resource users, scientists, government bodies and other stakeholders. Proper engagement between these groups will enable the development of management initiatives that are realistic and enforceable as well as effective for ecosystem management.

Embracing Change

Coupled social-ecological systems are constantly changing in ways that cannot be fully predicted or controlled. Understanding the resilience of ecosystems, i.e. the extent to which they can maintain structure, function, and identity in the face of disturbance, can enable better prediction of how ecosystems will respond to both natural and anthropogenic perturbations, and to changes in environmental management.

Multiple Objectives

Ecosystem-based management focuses on the diverse benefits provided by marine systems, rather than on single ecosystem services. Such benefits or services include vibrant commercial and recreational fisheries, biodiversity conservation, renewable energy from wind or waves and coastal protection.

Learning and Adaptation

Because of the lack of control and predictability of coupled social-ecological systems, an adaptive management approach is recommended.

Breakwaters: An Innovative Hard Option for Coastal

Breakwaters are structures constructed on coasts as part of coastal defense or to protect an anchorage from the effects of both weather and longshore drift.

***Figure:** Breakwaters create safer harbours, but can also trap sediment moving along the coast. Alamitos Bay, CA entrance channel.*

Purposes of Breakwaters

Breakwaters, also called bulkheads, reduce the intensity of wave action in inshore waters and thereby reduce coastal erosion or provide safe harbourage. Breakwaters may also be small structures designed to protect a gently sloping beach and placed one to three hundred feet offshore in relatively shallow water.

An anchorage is only safe if ships anchored there are protected from the force of high winds and powerful waves by some large underwater barrier which they can shelter behind. Natural harbours are formed by such barriers as headlands or reefs.

Artificial harbors can be created with the help of breakwaters. Mobile harbours, such as the D-Day Mulberry harbours, were floated into position and acted as breakwaters. Some natural harbours, such as those in Plymouth Sound, Portland Harbour and Cherbourg, have been enhanced or extended by breakwaters made of rock.

Unintended Consequences

The dissipation of energy and relative calm water created in the lee of the breakwaters often encourage accretion of sediment (as per the design of the breakwater scheme).

However this can lead to excessive salient build up, leading to tombolo formation reducing longshore drift shoreward of the breakwaters (Sea Palling, UK). This trapping of sediment can cause adverse effects down drift of the breakwaters leading to beach sediment starvation and increased erosion. This may then lead to further engineering protection being needed down drift of the breakwater development. Breakwaters are subject to damage, and overtopping in severe storms events.

Construction

Breakwaters can be constructed with one end linked to the shore, in which case they are usually classified as sea walls; otherwise they are positioned offshore from as little as 100 m up to 300-600 m from the original shoreline. There are two main types of offshore breakwater, single and multiple; single as the name suggests means the breakwater consists of one unbroken barrier, which multiple breakwaters (in numbers anywhere from 2-20) are positioned with gaps in between (50-300 m). Length of gap is largely governed by the interacting wavelengths. Breakwaters may be either fixed or floating, and impermeable or permeable to allow sediment transfer shoreward of the structures, the choice depending tidal range and water depth. They usually consist of large pieces of rock (granite) weighing up to 16 tonnes each, or rubble-mound. Their design is influenced by the angle of wave approach and other environmental parameters. Breakwater construction can be either parallel or perpendicular to the coast, depending on the shoreline requirements.

Types of Breakwater Structures

A breakwater structure is designed to absorb the energy of the waves that hit it, either by using mass (e.g., with caissons), or by using a revetment slope (e.g., with rock or concrete armour units).

In coastal engineering, a revetment is a land backed structure whilst a breakwater is a sea backed structure (i.e., water on both sides).

Caisson breakwaters typically have vertical sides and are usually erected where it is desirable to berth one or more vessels on the inner face of the breakwater. They use the mass of the caisson and the fill within it to resist the overturning forces applied by waves hitting them. They are relatively expensive to construct in shallow water, but in deeper sites they can offer a significant saving over revetment breakwaters.

Rubble mound breakwaters use structural voids to dissipate the wave energy. Rock or concrete armour units on the outside of the structure absorb most of the energy, while gravels or sands prevent the wave energy's continuing through the breakwater core. The slopes of the revetment are typically between 1:1 and 1:2, depending upon the materials used. In shallow water, revetment breakwaters are usually relatively inexpensive. As water depth increases, the material requirements, and hence costs, increase significantly.

Advanced Numerical Study

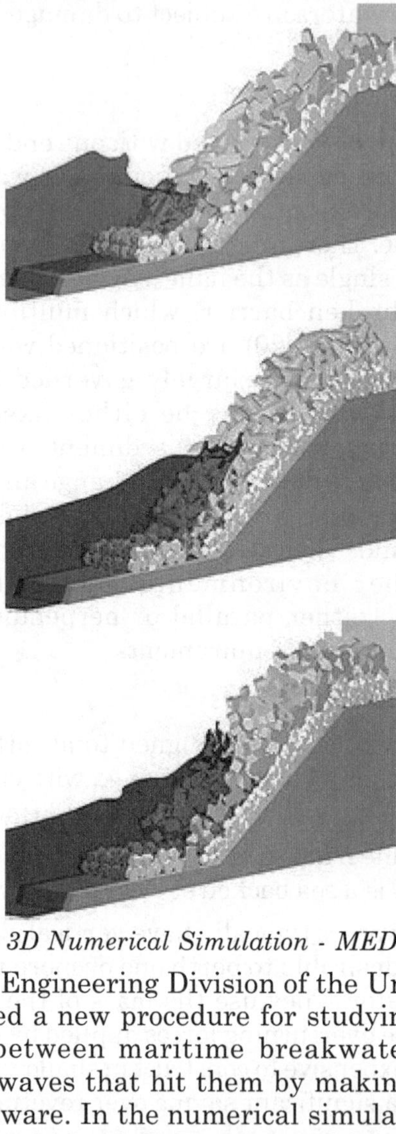

Figure: *3D Numerical Simulation - MEDUS 2009*

The Maritime Engineering Division of the University of Salerno (MEDUS) developed a new procedure for studying in greater detail the interactions between maritime breakwaters (submerged or emerged) and the waves that hit them by making integrated use of CAD and CFD software. In the numerical simulations, the filtration motion of the fluid within the interstices, which normally exist in a breakwater, is estimated by integrating the RANS equations, coupled with a RNG turbulence model inside the voids, instead of using classical equations for porous media. The breakwaters were modelled, in analogy to full size construction or physical laboratory tests, by overlapping three-dimensional elements and having the numerical grid thickened in order to have some computational nodes along the flow paths among the breakwater's blocks.

Bibliography

Asaithambi S. : *Economics of Ground Water Management in India*, Abhijeet, Delhi, 2008.

Bauer, Steffan: *Handbook of Globalization and the Environment*, CRC Press, Delhi, 2007.

Berlin, G.L.: *Earthquakes and the Urban Environment,* CRC Press, Boca Raton, FL, 1980.

Bhargava, Gopal: *Marine Ecosystem*, Kalpaz, Delhi, 2001.

Biswas, K P: *Corals of Tropical Oceans*, Daya, Delhi, 2008.

Boelaert-Suominen, S. and Cullinan, C.: *Legal and Institutional Aspects of Integrated Coastal Area Management in National Legislation*, FAO, 1994.

Bryant, D., E. Rodenburg: *Coastlines at Risk. An Index of Potential Development-Related Threats to Coastal Systems*, WRI, Washington, D.C., 1996.

Clark, J.: *Coastal Zone Management Handbook*, Lewis Publishers, New York, 1995.

Dolcemascolo, Glenn: *Environmental Degradation and Disaster Risk*, Asian Disaster Preparedness Centre, Sweden, 2004.

Drees, J.M.: *Coastal Dunes, Recreation and Planning. Proceedings of European Seminar*, EUCC, Leiden, The Netherlands, 1997.

Elmusa, Sharif S.: *Water Conflict: Economics, Politics, Law and Palestinian-Israeli Water Resources*, Institute for Palestine Studies, Beirut, Lebanon, 1998.

Engelman, R., and P. LeRoy : *Sustaining Water Washington*, D.C.: Population Services International, New York, 1993.

Epstein, S. J. M.: *The Earthy Soil: Bombay Peasants and the Indian Nationalist Movement, 1919-1947*, Oxford University Press, Delhi, 1988.

Fox, Lisa L.: *Management Strategies for Disaster Preparedness*, Chicago, American Library Association, 1989.

Garcia Nova, F., Crawford, R.M.M., Diaz Barradas: *The Ecology and Conservation of European Dunes*, Universidad de Sevilla, Spain, 1997.

Garrison, T.: *Oceanography, An Invitation to Marine Science*, Wadsworth, Belmont, 1993.

Hadfield, Peter: *Sixty Seconds that Will Change the World: The Coming Tokyo Earthquake*, Boston, C.E. Tutle, Co., 1992.

Hays, J.N.: *Epidemics and Pandemics: Their Impacts on Human History*, Santa Barbara, Calif.: ABC-CLIO, 2005.

Healy M.G. and Doody, J.P.: *Directions in European Coastal Management*, EUCC, Samara Publishing, Cardigan, 1995.

Jeftic, L., Milliman, J., Sestini, G.: *Climate Change and the Mediterranean—Environmental and Societal Impacts of Climatic Change and Sea-Level Rise in the Mediterranean Region*, Edward Arnold, London, UK, 1992.

Jones, P.S., Healy M.G. and Williams, A.T.: *Studies in European Coastal Management*, EUCC, Samara Publishing, Cardigan, 1996.

Kaur, Kanwaljit: *Handbook of Water and Wastewater Analysis*, Atlantic, Delhi, 2007.

Lewis, Peirce. *New Orleans: The Making of an Urban Landscape*. Cambridge, MA: Ballinger Publishing Co., 1976. Annotation

Merritt, John F. : *History at Risk, Loma Prieta: Seismic Safety & Historic Buildings*, California Preservation Foundation, 1990.

Miller, H. and S.-L. Shaw: *Geographic Information Systems for Transportation: Principals and Applications*. New York: Oxford, 2001 University Press.

Nash, J.M.: *El Nino: Unlocking the Secrets of the Master Weather Maker*, Warner, Delhi, 2002.

Odum, Howard T.: *Environment, Power and Society*, New York, Wiley-Interscience, 1971.

Oliver, John E.: *Encyclopedia of world climatology*, Springer, Delhi, 2005.

Pernetta, J., and Elder, D.: *Cross-sectoral, Integrated Coastal Planning: Guidelines and Principles for Coastal Area Development*, Gland, Switzerland, IUCN, 1993.

Pilarcyzk, K.W.: *Coastal Protection*, Balkema, Rotterdam, The Netherlands, 1990.

Salman, A., Berends H., Bonazountas M.: *Coastal Management and Habitat Conservation*, Marathon, Greece, April, 1993. EUCC, Leiden, 1995.

Stowe, K.: *Ocean Science,* Wiley, New York, 1983.

Vallega, A.: *Fundamentals of Integrated Coastal Management*, Kluwer Academic Publishers, UK, 1999.

Varley, Anne: *Disaster, Development Environments*, New York, J. Wiley, 1994.

Welp, M.: *Defining Integrated Coastal Zone Management for the Baltic Sea Region*, Springer-Verlag, Berlin, Heidelberg, 1999.

Index

I

L

M

N

O

P

R

S

T

U

V

W

Z

❑❑❑